1797

. . . the island of Rum will one day be considered, if not the most remarkable of the Hebrides, at least a very important field of enquiry.

EDWARD DANIEL CLARKE

1824

The interior [of Rhum] is one heap of rude mountains, scarcely possessing an acre of level land. It is the wildest and most repulsive of all the islands . . . If it is not always bad weather in Rhum it cannot be good very often.

JOHN MACCULLOCH

RHUM

The Natural History
of an Island

⌣⌣⌣⌣⌣⌣⌣⌣⌣⌣⌣⌣⌣⌣⌣⌣⌣⌣⌣⌣

EDITED BY
T. H. CLUTTON-BROCK AND M. E. BALL
FOR THE EDINBURGH
UNIVERSITY
PRESS

© T. Clutton-Brock and M. E. Ball 1987

Edinburgh University Press
22 George Square, Edinburgh

Set in Linotronic Plantin by
Speedspools, Edinburgh, and
printed in Great Britain by
Lindsay & Co. Ltd, Edinburgh

British Library Cataloguing
  in Publication Data
Rhum: the natural history of an
  island—(Island biology series)
1. Natural history—Scotland—
  Rum, Isle of
I. Clutton-Brock, Tim   II. Ball, M. E.
574.9411′85   QH141
ISBN 0 85224 513 0 (cloth)
     0 85224 570 X (paper)

CONTENTS

FOREWORD

J. MORTON BOYD CBE

Many people have spoken of Rhum as the 'forbidden island', but they are strangers; others who have been there speak of the 'Rhum experience' – a personal response, deeply felt and long remembered. There are two sides to it; one is the ever-fresh feeling of being again among island friends in the enjoyment of a little world apart and the other is the spell which the island itself casts upon islander and visitor alike.

The welcome to Rhum comes from the quietly-spoken crew of the island ferry-boat *Rhouma*, gently man-handling people and baggage to a safe landing at the Kinloch boat-slip. In my visits over the years as the Director in Scotland of the Nature Conservancy Council who own Rhum, I had things to see in many parts of the island. I prided myself in my contact with such fine country and the people who managed it; I always set aside time to be on the hill with the Chief Warden. These hill-walks with Peter Wormell, George Mac-Naughton, Peter Corkhill, Bob Sutton and Laughton Johnston, in their successive tours of duty, were among the finest I have known. The companionship when out to Askival or Bloodstone Hill, the unfolding splendour of the island in changing moods of weather, the thrill of seeing wildlife in the grand setting, the magic of sunbeams upon mountain and sea, all combined to quicken our minds and hearts; we talked often of a *Book of Rhum*.

One of my favourite days on Rhum is a circular walk around the Cuillin starting at Kinloch, over to Harris by road, past Loch Fiachanis (Sand Loch) ascending the narrow defile to Bealach an Fhuarain (Pass of the Spring), down to the bothy at Dibidil and back to Kinloch over Welshman's Rock. If the day is hot the bottle-green cisterns of the burns at Harris, Dibidil and Allt na h-Uamha are a delight for a plunge. The high point of the walk is on the Bealach which, set between the impending walls of Trollaval and Ainshval, is an enthralling place – the heart of Rhum. It is Vulcan's forge in the cold, eroded core of the ancient volcano which is cast in a great whorl of igneous peaks and precipices. The venturi effect of the narrowing defile of Fiachanis accelerates the wind, and sends it roaring against the naked flanks of Trollaval. Dark clouds form from which croaking ravens appear and disappear. After dark, nesting shearwaters fly to their tenemented city high in the Cuillin, their staccato calls echoing eerily in the pass.

The vision of Rhum possessed by the founding fathers of nature conservation was the great outdoor laboratory and demonstration area in the Scottish Highlands and Islands, stemming from an original idea of Frank Fraser Darling. It might never have become a reality but in 1957, when the far-sighted Max Nicholson was Director General of the Nature Convervancy, the island was purchased by the Conservancy, together with Kinloch Castle (in working order) and all other buildings except the Mausoleum at Harris which was retained by the Bullough family, the previous owners. The story since then has two main strands; one is contained in this book and covers the scientific effort which has been expended on Rhum over the last thirty years; the other is contained in the files of the Nature Convervancy Council (NCC) and covers the effort in organisation, administration and management without which the continuous scientific work would not have been possible. The foundations were laid by Joe Eggeling who wrote the first management plan in 1960; John Arbuthnott, who set up the estate; Peter Wormell and George MacNaughton, who made it work on the ground in the 1960s. Since then many others have continued the building up of Rhum as one of the most prestigious nature reserves in Europe.

The management plan has struck a balance between man's several activities on the island towards the fulfilment of research objectives in the wild bounds of Rhum. This book, therefore, is testimony not only to the great amount of research which has been done but also to how well the plan has worked. Today, the island could not be more celebrated as a conservation area; it carries the treble accolade of National

Scenic Area, National Nature Reserve, Biosphere Reserve (under the 'Man and Biosphere' programme of UNESCO) and in future it could become one of the Britain's sites listed under the World Heritage Convention.

This book has brought together a great range of work in the natural and human history of Rhum much of which is published in scientific journals and books unavailable to the general reader. The authors have invested a total of some eighty man-years of interest in Rhum, they possess a most formidable body of knowledge of the island and have made major contributions to its scientific record and literature. Here they collaborate in a single more popular work which serves the double purpose of reviewing the progress of research and providing a description of the island as a whole. Though each chapter deals with a separate facet of Rhum's character, the whole book forms a detailed account of the ecosystem of a small, mountainous, temperate, maritime island, known widely in nature conservation.

The book reaffirms the purpose of the Rhum National Nature Reserve based on high scientific ideals and endeavours and set upon a steady course of continuous evolution of the environment towards stated goals in nature conservation. The future holds the regrowth of the forest, the management of red deer in a changing range, the enhancement and redistribution of wildlife, the use of livestock as 'tools' in wildlife management and the opportunity to experiment with different regimes of natural and semi-natural vegetation, herbivores and predators. All this exciting vista is only possible, however, if Rhum continues to be a well-managed, researched and monitored nature reserve.

Rhum is a symbol of the inspired optimism of the 1950s for the future of nature conservation in the Highlands and Islands of Scotland. In the 1930s no-one would have believed that such an island would be held in trust nationally for the sole purpose of making it a nature reserve; now, in the 1980s, the optimism of the 'fifties has waned and the NCC is hard-pressed to retain the reserve in its former condition. However, the Wildlife and Countryside Act has placed nature conservation in Britain on a much firmer footing than ever before. More funds are being made available to the NCC and, though there are many competing claims, hopes are high for Rhum and for further well-managed reserves in future.

Rhum is also symbolic of a society conscious of its achievements in civilisation and enlightenment. Its survival as a world-class nature reserve will only be possible, however, if the Nature Conservancy Council (or other similar national conservation body of proven worth) continues to own and manage the island. Thus the true value of Rhum as an enduring experiment between man and wildlife will be realised for all time.

1

INTRODUCTION

T. H. CLUTTON-BROCK AND M. E. BALL

Rhum is one of Britain's largest National Nature Reserves – an island owned by the Nation and set aside by Act of Parliament for the purposes of conservation, research and education. It is one of the Inner Hebrides – a chain of 22 islands running along the western seaboard of Scotland from the northern tip of Skye (57°41′N) to the southern point of Islay (55°34′N) which range in size from a few hectares to the large islands of Skye and Mull. Rhum lies 12 km south-west of Skye and is the largest of the Small Isles – a group of five islands including Canna, Sanday, Eigg and Muck (figure 1.1).

Of the Small Isles, Rhum is the most mountainous. Volcanic activity during the Tertiary era disrupted the far older sandstones that extended over much of the west coast of Scotland and threw up the island's main hills – of which Askival (812 m), Hallival (723 m), Barkeval (591 m), Trollval (or Trallval) and Ainshval (702 m) survive as eroded remnants (figure 1.2). A major fault separates these jagged peaks from the flatter hills of sandstone granite or basalt found in the north and west of the island.

During the Ice Age, Rhum's hills were heavily glaciated by their own ice cap which scored their flanks, leaving numerous corries, moraines and drift deposits. As the ice finally withdrew some 15,000 years ago, meltwater streams cut beds in the glen bottoms, falling steeply towards the sea. Today, they still flow quickly through rocky ravines, except the Kilmory River which rises in the volcanic, ultrabasic rocks of the south east and drains to the north through alluvial marshes and sandy machair before meeting the sea at Kilmory Bay. The island has numerous hill lochs and lochans

Figure 1.1. The west coast of Scotland, showing Skye and the Inner Hebrides. Rhum lies to the south of Skye at 57°N 6°20′W.

– mostly shallow and peaty. The only lowland loch is at Papadil in the south east – a shallow basin below the steep slopes of Sgurr nan Gillean, separated from the sea by a narrow strip of beach.

Today, the climate is cool and wet with average rainfall of 1800 mm a year in the coastal zone and 3000 mm in the hills. April or May are the driest months and October, November or January are usually the wettest – but years vary

2

Figure 1.2. 'The interior is one heap of rude mountains, scarcely possessing an acre of level land . . .'. Looking across Sgurr nan Gillean to Ainshval with Askival, Hallival and Kinloch behind.

widely and any month of the year can be the wettest. Ground frost is common between October and May but, except on the highest peaks, snow rarely lies for long and the coastal zone is covered for less than thirty days in most years.

The island's wet climate has favoured the formation of blanket peat on all but the steepest slopes and best drained soils where small patches of brown earths or podsolized soils occur (figure 1.3). On the upper slopes of the ultrabasic peaks which are too steep to allow peat to form, deep brown earths are found over large areas, associated with colonies of manx shearwaters. Over the centuries, the burrows and droppings of these birds have encouraged the development of a closed turf of grasses and herbs and prevented erosion.

After the disappearance of the ice cap, the island was

Figure 1.3. Peat beds now cover the glen bottoms. Looking across Kilmory Glen to Mullach Mor.

colonised by sedges, grasses, herbs, shrubs and trees. Birch and willow, wych elm, oak, hazel and alder, holly, rowan and Scots pine grew on the lower slopes of the hills and in the glen bottoms. With them came a diverse vertebrate and invertebrate fauna. Sea bird colonies formed on the steep cliffs, golden plover and red grouse nested on the hill slopes, golden and sea eagles bred among the peaks.

Mesolithic scrapers of around 6600 BC provide the first records of human activity and from later periods there are stone arrowheads. Such finds indicate that the island has been inhabited by man since the seventh millenium BC. Later, Norsemen named the prominent hills and valleys that may have served as landmarks to the longships – Askival, Hallival, Guirdil and Dibidil among them. The island's name may derive from the Norse word 'Romoy' meaning wide or spacious island, but the bulk of the island's population was of Celtic origin, as the Gaelic names for the lochs, streams and gulleys show.

Norse domination of the Inner Hebrides continued until 1156 when Somerled of Argyll, a petty king of mixed Norse and Celtic blood, won control of the southern islands. However, he could not maintain this for long and during the Middle Ages the Small Isles were the scene of regular local warfare. Short periods of stability were interspersed with long bouts of lawlessness. In the sixteenth century, the Small Isles came under the control of the Scottish crown. Rhum's human population was over three hundred by the end of the eighteenth century – mostly in the settlements at Harris and Kilmory (see figure 1.4). Felling, burning and overgrazing by sheep cleared the extensive woodland and the red deer were exterminated. Finally, in 1826, after a change in the ownership of the island and a rise in wool prices, the landlord evicted the island's inhabitants, replacing them with eight thousand blackface sheep, but the prices of wool and mutton fell and the venture quickly proved profitless.

In 1845, the island was bought by the second Marquis of Salisbury, red deer were reintroduced, and it was managed as a sporting estate and, secondarily, as sheep grazing. During the nineteenth century, it passed through the hands of other owners until, in 1887, it was purchased by John Bullough, a Lancashire cotton baron. His son, George, a world traveller and friend of royalty, inherited the island and built

Figure 1.4. The ruins of Kilmory village. Looking south along Kilmory Glen to Hallival and Askival.

Figure 1.5. The castle, built by the Bulloughs at Kinloch, of red sandstone imported from Arran.

shooting lodges at Harris and Papadil and a baronial castle at Kinloch (figure 1.5). After the First World War the estate began to decline and Rhum was finally sold to the Nature Conservancy in 1957.

Among Highland estates, the history of Rhum is not unusual. Throughout much of the Western Highlands humans and domestic animals cleared much of the forested ground between the sixteenth and eighteenth centuries before being evicted to make way for sheep and deer. Time has healed the scars on the landscape and, today, it is easy to forget that the heather- and *Molinia*-dominated moorland that covers so much of the Scottish west coast is a man-made desert, created by human interference.

The most distinctive phase in Rhum's history is the most recent one – for of the 500-odd deer forests in Scotland only a handful have become National Nature Reserves. In some ways, the island was a surprising choice. Apart from the mountain colony of Manx shearwaters, it harboured no animal populations that are not reasonably well represented elsewhere. Its peaty acres are mostly covered with the same moorland plants that are found across upland Scotland. And though its geology is unique and the ledges of its southern hills carry a number of uncommon alpine plants there are many other areas of the Scottish mainland whose rarities require more urgent protection.

The far-sighted members of the Nature Conservancy who purchased Rhum in 1957 realised that its importance lay in a different direction – as an ideal site for research, experimental management and education. Rhum includes a wide range of habitats representative of the north west of Scotland. To answer many of the most important questions about moorland ecosystems and to investigate novel techniques of management, a site was needed that was not subject to the constraints of private ownership. In particular, the Nature Conservancy wished to determine whether or not it was possible to establish native tree species on the acid peaty soils and to recreate the natural woodland stripped from the countryside over two hundred years ago. In addition, Rhum's open habitat, finite boundaries and isolation from human interference made it an ideal site for research on the ecology of moorland animals, especially red deer.

The first management plan drawn up for Rhum in the

early sixties outlined the Nature Conservancy's plans – to recreate the natural characteristics of a Hebridean island by restoring productivity and species diversity and to use the island as a base for research and education. Sheep were removed and the practice of burning the moorland vegetation (muirburn) each spring to encourage new growth was stopped. The island's staff were to include a chief warden, in overall charge; a warden to be responsible for the management of the island and the day to day activities of the staff; a stockman to look after a small dairy herd as well as the ponies used for recovering deer carcasses; a ferryman; three estate workers; and the castle caretakers. In the early years of the Nature Conservancy's ownership, priority was given to research on the geology, soils and climate, and to cataloguing the plant and animal species found on the island and to mapping the vegetation communities. Research on the ecology of the red deer population was started, a tree nursery was established at Kinloch and the task of replanting the island with native tree species began.

By the early seventies, most of the necessary censuses had been completed, the research on red deer had produced a quantitative description of the life history of the species which provided a sound basis for deer management throughout Scotland and trial woodland plantations around Loch Scresort and elsewhere on the island were well established. Rhum's potential as a site for field research was realised by several university-based scientists and independent projects began to examine the geology of the island, the behaviour and reproductive physiology of the deer and the ecology of the vast shearwater populations breeding on Askival and Hallival. In addition, the growing woodlands stimulated research on the establishment and growth of bird populations and, in 1975, the Nature Conservancy Council decided to use the island as a base for reintroducing the sea eagle to the British Isles. During the same period, a free-ranging herd of Highland cattle was established at Harris in an attempt to redress changes in the vegetation that had arisen following the removal of sheep and the reduction in grazing pressure after 1957.

In the late seventies and early eighties, financial constraints led to reductions in the island's staff from eleven families to a maximum of eight and a minimum of six, whose

8

children are taught by a resident school teacher. Replanting continues on the south side of Loch Scresort, now assisted by the British Trust for Conservation. Volunteers plant 25,000 trees and shrubs of nineteen different species each year. Sea eagles are established on Rhum and in other parts of Scotland and have made several attempts to breed, while long-term research on red deer population regulation and reproductive success continues in the North Block of the island, based at Kilmory. Some 4,000 visitors land annually on Rhum and the castle provides both hotel and hostel accommodation.

This book summarises some of the studies that have been carried out on Rhum during the last twenty-five years. Henry Emeleus, Reader in Geology at Durham, synthesizes knowledge of the volcanic events that established the island's physical structure. John Love, the scientist in charge of the project to re-establish sea eagles on the island and an amateur historian, outlines the human history of Rhum as well as contributing to a chapter on its bird populations. Martin Ball, the Nature Conservancy officer responsible for Rhum describes the island's flora and tells the history of the attempts to reforest part of the island. Peter Wormell, the Chief Warden of Rhum from 1958 until 1973, summarizes his studies of the island's invertebrates and land birds. Tim Clutton-Brock of Cambridge University and Fiona Guinness outline the findings of the research on the island's red deer population, which began in 1958 and is still thriving today, while Iain Gordon, Robin Dunbar, David Buckland and David Miller (also from Cambridge) describe studies of the island's hill ponies, Highland cattle and feral goats.

The extent and detail of these studies have made Rhum one of the best documented field sites in Great Britain. The breadth of past studies offers future scientists the possibility of working on a comparatively simple ecosystem whose component species have already been documented. More important, the long span of routine records of changes in climate, vegetation and animal numbers, now extending over more than twenty years, allows investigation of long-term interactions between plant and animal communities. Past research has clearly shown that many of the most important ecological changes occur on a timespan that must be measured in decades rather than years and the importance of the island's contribution to our understanding of ecological processes will grow as the years pass.

Further Reading

Nature Conservancy Council (1974) *Isle of Rhum National Reserve, Reserve Handbook*. Nature Conservancy Council, Inverness.

Eggeling, W. J. (1964). A nature reserve management plan for the Isle of Rhum, *J. Appl. Ecol. 1*, 405-19.

2

THE RHUM VOLCANO

C.H.EMELEUS

Introduction

Sixty million years ago, western Europe was close to Greenland, and the north Atlantic was restricted to a narrow strait
near Rockall. Over the next ten million years or so, in the
early Tertiary, or Palaeocene (table 2.1), Greenland and
Europe separated further, the north Atlantic developed and
extensive volcanic activity associated with these events occurred in several areas, including the north-western British
Isles. In Scotland, northwest – southeast fissures were produced by the movement of land masses. Lavas of basaltic
composition flowed from these and covered and drowned the
Palaeocene land surfaces. Today, these form a thick succession of lava flows on Mull, northern Skye and parts of the
Small Isles, including Rhum. However, not all the basaltic
liquid (generally termed basaltic magma) reached the surface, much of it solidified in the fissures and today forms
'swarms' of near-vertical, wall-like dykes of basalt. These
swarms are extremely well preserved along the Rhum coast
between Kilmory and Guirdil and on the adjoining islands,
especially Muck.

The widespread lava fields and dyke swarms were followed at several places by the formation of large volcanoes.
The mountainous southern part of Rhum (figure 2.1), the
Skye Cuillin and Red Hills, and the mountains of central
Mull and the Ardnamurchan peninsula all mark sites of
extinct Tertiary volcanoes. Erosion of the volcanoes started
even as they were formed and continues to the present day,
though it was particularly rapid in the Pleistocene (table
2.1). The upper parts of the volcanoes and the upper parts of

Table 2.1. Rhum and the geological time scale, showing the start of the periods represented on Rhum in millions of years before present.

Eras and Periods		Start of Period	Evidence
Quaternary	Recent	0.01	Peat, river alluvium, blown sand on north coast, etc.
	Pleistocene	1.5-2	Corrie glaciation, moraines, raised beach deposits, glaciation by mainland ice.
Tertiary	Palaeocene	65	Lavas, sediments of NW Rhum. Erosional interval. Ultrabasic rocks and gabbros. Main Ring Fault, with central uplift. Major part of NW/SE dyke swarm.
			Granophyres, microgranites, porphyritic felsite, explosion breccias, early gabbroic plugs. ? early lavas.
Mesozoic	Jurassic	190	Altered limestones at Allt nam Ba
	Triassic	235	Coarse sandstones, etc., with basal limestone (cornstones) of Monadh Dubh, NW Rhum.
Palaeozoic			Small thrust fault at Welshman's Rock
Precambrian	Torridonian	950	Extensive feldspathic sandstones over eastern Rhum.
	Lewisian	*c.*2500	Gneisses at the Priomh Lochs and near Papadil.

the lava fields are now mostly missing. In the volcanoes, erosion has cut down to their roots, exposing the great variety of rock types and structures involved in the volcanic activity. On Rhum, erosion has been so severe that at least one kilometre of cover has been removed since the early Tertiary.

The Pre-Tertiary Framework

The Rhum Tertiary volcano developed on a ridge of ancient Precambrian rocks flanked by basin-like structures filled with Mesozoic sediments (table 2.1). Before the volcanic activity of the Tertiary, Rhum was made principally of a thick succession of Torridonian sandstones and shales resting on Lewisian gneiss. About 3000 m of Torridonian strata are now preserved, which are inclined to the west-north-west at angles of between 15°–25°. These weather unevenly to provide the prominent inclined benches visible on the north side of Kinloch Glen and as far west as Guirdil. These Torridonian strata are found to lie on an ancient land surface eroded from Lewisian gneiss near the Priomh Lochs, east of the Long Loch (figure 2.1).

Younger than the Lewisian and Torridonian rocks, but older than the Tertiary lava flows are the brown and orange sandstones north of Glen Shellesder, on Monadh Dubh. At their base a prominent, white layer of limestone ('cornstone') has weathered to give lime-rich soils which support the unusual plants of this area. These Triassic rocks (table 2.1) were formed by the weathering of Torridonian sandstones under semi-desert conditions. In addition, a small area of sandstone, limestone and shale of younger age than the Triassic strata but also older than the Tertiary lavas crops out in eastern Rhum, on the north-east slopes of Beinn nan Stac, south of Allt nam Ba (figure 2.1). These rocks, which were altered and baked by the Tertiary volcano, contain many fossils and are of similar age to Jurassic strata found on Skye.

Growth and Decline of the Tertiary Volcano

The evolution of the Rhum volcano and the formation of Rhum's Tertiary igneous rocks (table 2.2) fall into three fairly well-defined stages. First, there was a period when granitic magmas were dominant, although lesser amounts of basaltic magma were present. This phase was accompanied by ring-faulting producing the marked break visible between the Torridonian strata of northern and eastern Rhum and the rocks of the volcanic centre. A second period of igneous activity followed when the peridotites, allivalites and gabbros of Barkeval, Hallival, Askival and Trallval were formed. In the third, final stage, erosion excavated the volcano to its roots, deep valleys were carved on the flanks and were filled by debris stripped off the volcano as well as by flows of basaltic lava from sources outside Rhum.

Early Growth of the Volcano

An unusual feature of the Rhum volcano is the apparent absence of early lava flows. The only evidence for their

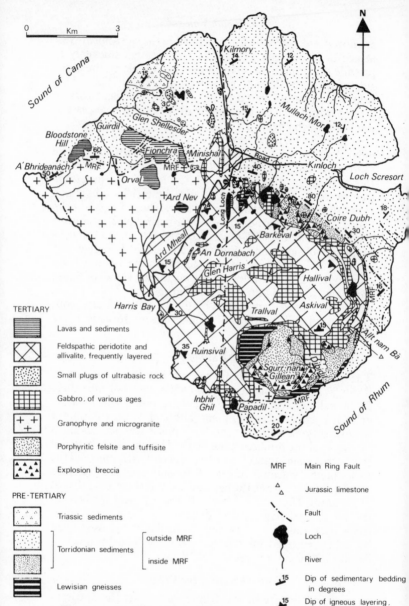

TERTIARY

Lavas and sediments
Feldspathic peridotite and allivalite, frequently layered
Small plugs of ultrabasic rock
Gabbro, of various ages
Granophyre and microgranite
Porphyritic felsite and tuffisite
Explosion breccia

PRE-TERTIARY

Triassic sediments
Torridonian sediments { outside MRF / inside MRF }
Lewisian gneisses

MRF — Main Ring Fault

△△ — Jurassic limestone

— Fault

⬤ — Loch

— River

↘15 — Dip of sedimentary bedding in degrees

▲15 — Dip of igneous layering, in degrees

Figure 2.1. Geological map of Rhum.

14

former presence comes from basaltic rocks close to the Jurassic limestones south of Allt nam Ba and from baked basaltic inclusions ('xenoliths') within parts of the allivalites east of Hallival. Early lava flows do occur on Eigg and Muck and it is likely that these once extended over Rhum but were removed by erosion. The remains of the conduits from which the lavas flowed can be seen in the small gabbroic bodies (or 'plugs') which are particularly abundant piercing the Torridonian strata on the southern side of Kinloch Glen (figure 2.1).

South of Kinloch Glen and north of the high hills of southern Rhum there are three small rounded hills, Cnapan Breaca and Meall Breac either side of Coire Dubh, and Am Mam further to the west. These are made of felsite, a fine-grained rock of granitic composition (table 2.2), and were formed when magma in volcanic vents solidified very rapidly as it came close to the Tertiary land surface. The uprise of the magma was accompanied by violent explosions as evidenced by the shattered and broken state of Torridonian sandstones and Lewisian gneisses adjoining the felsite. Good examples of these fragmented rocks, termed 'explosion breccias', are visible close to the stream on the floor of Coire Dubh (figure 2.1). Similar felsite intrusion and explosive activity also occurred in southern Rhum where the rocks are well displayed on the south-west side of Dibidil, on Sgurr nan Gillean and as far west as Ainshval.

Granite was also formed during the growth of the volcano. On Rhum this rock is rather finer-grained than the granite found at Strontian, in the Cairngorms and at other places in Scotland; the variety present on Rhum is more correctly termed 'microgranite' (figure 2.1, table 2.2). The microgranite forms the smooth, vegetation and scree-covered hills of Ard Nev and Orval and it is also magnificently exposed in the sea cliffs from Harris Bay to A' Bhrideanach in western Rhum. Unlike the felsites, the microgranite solidified beneath a cover of older rocks. The remains of this roof are visible on the summit of Ard Nev where baked Lewisian gneiss rests on microgranite.

Although the early stages of the Rhum volcano were dominated by the intrusion of granitic magmas, basaltic intrusions preceded these and there is also evidence of the mixing of granite and basalt magmas on slabs high on the west side of Dibidil. Basaltic magmas continued to be in-truded after the granite magma ceased to rise for there are numerous examples where both granite and felsite are cut by basaltic dykes: networks of basaltic dykes intrude felsite on the east of Cnapan Breaca and also on the slabs and crags of western Dibidil; in western Rhum, many dark basaltic dykes are visible in the granite on the west side of Harris Bay and in the deep clefts in the granite along the base of the cliffs further to the north-west.

The geological map of Rhum (figure 2.1) reveals an odd feature shared by the majority of the igneous rocks described so far: they have faulted boundaries against many of the earlier rocks. In particular, neither granite nor felsite in-trudes into the Torridonian strata to the north or east. The vertical or steeply-inclined character of these faulted bound-aries is clear: the steep boundary between granite and Tor-ridonian sediments is visible in the gullies on the west of Guirdil, about one km south-east of Bloodstone Hill; simi-larly, a near-vertical, curving fault circumscribes the vol-canic rocks between the Long Loch and Cnapan Breaca. Dark, flinty shattered rocks ('mylonites') along the fault are present near the deer fence gate in Coire Dubh. The signifi-cance of these faults was realised by Sir Edward Bailey, a former Director of the Geological Survey, when he visited the island about forty years ago. Bailey showed that the faults were part of a major ring fault which he termed the Main Ring Fault (figure 2.1). However, unlike the ring faults of Mull, Ardnamurchan and other Scottish Tertiary volcanic centres, Bailey demonstrated that on Rhum there had been considerable *upwards* displacement of many of the rocks within the Main Ring Fault, possibly by as much as 2000 m. The crucial evidence for this unusual situation comes from the restriction of both the Lewisian gneisses and the lower-most (and therefore oldest) Torridonian sediments to within the Main Ring Fault: nowhere on Rhum do these rocks occur outside this fault although they must be present a few hundreds of metres beneath the present-day surface of east-ern Rhum. Research on Rhum since Bailey's investigations has revealed an even more complex history of movement. The uplift recognised by Bailey within the Main Ring Fault has been confirmed, but it was followed by equally profound *subsidence* within the fault, bringing down the Jurassic limestones which crop out south of Allt nam Ba (figure 2.1). This movement was followed by further substantial uplift within the Main Ring Fault. Clearly, the block of microgranite, felsite, explosion breccia, Torridonian sediments, Lewisian gneisses and other rocks bounded by the Main Ring Fault has behaved like a piston and it is probable that the power for this movement was provided by granitic magmas rising from deep within the Rhum volcano.

The Main Ring Fault may be examined at the places mentioned above and it is also visible in a line of low scarps and cliffs crossing over the south-east side of Beinn nan Stac from Allt nam Ba to the sea cliffs on the north side of Dibidil. However, the most obvious structures caused by the fault are the folding and large-scale distortion of bedding in the Torridonian sandstones and shales. In western Rhum, between Bloodstone Hill and A' Bhrideanach, the steeply inclined slabs of sandstone in the cliffs of Sgorr Mhor are bedding planes tilted up against the faulted contact with the microgranite; similarly, the steep slabs inclined towards the road all along the south side of Kinloch Glen owe their attitude to tilting against the Main Ring Fault.

The Culmination of the Volcano

The core of the Rhum volcano consists of peridotites and allivalites (collectively termed 'ultrabasic rocks'), (table 2.2), and gabbros which form much of the spectacular scenery on the island including the hills of Askival, Hallival, Barkeval, Trallval and Ruinsival. The outstanding feature of the ultrabasic rocks, for which Rhum is known world-wide, is their spectacular bedded or layered structure (figure 2.2). The resistant rock forming the lines of scarps and cliffs is allivalite, and the slack ground between results from the easy weathering of peridotite which has eroded to give the boulder-strewn terraces favoured as nesting grounds by the Manx shearwaters. The origins of the layering of the ultrabasic rocks are still the subject of much research. However, many geologists believe the theory advanced by G.M. Brown, namely that the layered rocks formed through the differential settling, or sedimentation, of crystals precipitating from basaltic magmas undergoing very slow cooling. Each layer consists of a lower part made of peridotite, a rock rich in the dense mineral olivine, which passes upwards into allivalite, a

Table 2.2. The igneous rocks of Rhum and their minerals.

Basalt: a dark, fine-grained rock (crystals $<$ 1 mm) consisting of pyroxene ($CaSiO_3.(Mg,Fe)SiO_3$), plagioclase feldspar ($CaAl_2Si_2O_8$) and iron-titanium oxides, sometimes accompanied by olivine (($(Mg,Fe)_2SiO_4$). Small gas cavities are often present, which may contain agate (SiO_2), and lava flows on Bloodstone Hill and western Fionchra have cavities and small fissures filled by bloodstone.

Gabbro: a dark, fairly dense rock of similar mineralogy to basalt but coarser grained, with crystals generally 4-5 mm or more in diameter.

Ultrabasic rocks: these are somewhat similar to gabbro in their mineralogy and grain sizes except that the olivine and pyroxene have higher Mg/Fe ratios, and the plagioclase feldspar has a higher Ca/Na ratio. Two varieties are common on Rhum:

i) Peridotite: a dense rock, deep brown to warm brown on the weathered surface but dull black on freshly broken surfaces. The rock is extremely rich in olivine but as plagioclase is also present the Rhum rocks should strictly be termed 'feldspathic peridotites' (figure 2.1). The olivine may weather to give small pits on the surface of the rock; feldspar tends to stand proud and may form lace-like patterns on weathered surfaces.

ii) Allivalite: this is a medium to pale grey rock containing olivine and plagioclase in approximately equal amounts, together with a lesser proportion of pyroxene. The tabular plagioclase crystals may show a strong parallel alignment, giving the rock a 'laminated' structure.

The mineral chromite ($FeO.Cr_2O_3$) occurs in the ultrabasic rocks. Concentrations of chromite are often visible at the boundaries between allivalite and the peridotite of the next overlying layer. The concentrations are in the form of jet black seams a few mm in thickness.

Microgranite: light cream or white rock of relatively low density, with crystals from 1-3 mm in length. The principal minerals are quartz (SiO_2), sodium-rich plagioclase feldspar and orthoclase feldspar ($KAlSi_3O_8$). Scattered small, black, shining flakes of biotite mica ($K_2(Mg,Fe)_2(OH)_2(AlSi_3O_{10})$) are also present. Small gas cavities ('drusy cavities') are sometimes visible, which may be lined with well-crystallised quartz and feldspars. (Granophyre (figure 2.1) is a rather finer-grained variety with a distinctive microscopic texture.)

Felsite: this is similar in chemical composition to microgranite but consists of small (1-2 mm in length) well-formed crystals of white plagioclase, grey quartz and black pyroxene set in a very fine-grained (crystals ≪ 1 mm), dark grey to black matrix. 'Porphyritic felsite' (figure 2.1) refers to the texture of the rock, which has well-formed crystals in a much finer-grained matrix. The term 'tuffisite' (figure 2.1) refers to felsite which has been shattered by gas explosions in the volcano.

---

light-coloured rock with a high proportion (about fifty per cent) of the less-dense mineral plagioclase (table 2.2). At least fifteen of these rhythmically repeated layers, varying in thickness from 50 to 150 m, are present in eastern Rhum, on and around Hallival and Askival. Each layer is thought to represent a magmatic sediment formed by the settling of olivine, plagioclase feldspar and other minerals on the floor of a very fluid pool of basaltic magma deep within the volcano. The denser, and probably early-formed olivines were first to settle, to be joined by the lighter feldspar which eventually became the dominant mineral. Not all the basaltic magma crystallised at this stage. The residue is thought to have been expelled as surface flows from the volcano, expulsion occurring when a further batch of fresh magma rose from deeper levels in the volcano. The newly-arrived magma in turn cooled and crystallised, forming another, overlying layer. This sequence of events was repeated at least fifteen times in eastern Rhum.

Even to the casual observer, the similarities between the layering of the ultrabasic rocks and the bedding of the Torridonian are fairly apparent. The comparison is not superficial and there are many small-scale structures in the ultrabasic rocks which closely simulate those found in sediments such as sandstones deposited in water. Structures comparable with those of water-logged sediments occur where peridotite blocks appear to have subsided onto partly crystallised allivalite, bending and deforming fine-scale layering, and where complex, small-scale folding (figure 2.3) resembles deformation found in shales and sandstones which have slumped down gently-inclined slopes.

The ultrabasic rocks, and especially the peridotites, are appreciably denser at about 3.2 g/cm³ than most other rocks (which are between 2.65 and 2.9 g/cm³). This property has

Figure 2.2. Layered allivalites and feldspathic peridotites of the Eastern Layered Series forming the summit of Hallival. The layers of resistant, light-weathering allivalite stand out as lines of crags, the feldspathic peridotite weathers away to give the intervening shelves. Note also the westward-dipping Torridonian sandstones north of Kinloch Glen, *left*.

Figure 2.3. Slump structures in allivalite near the summit of Askival. (Scale: the hammer shaft is approx. 40 cm long.)

Figure 2.4. Intrusion breccia developed at contact of the granophyre (pale) and feldspathic peridotite, east side of Harris Bay. Blocks of basalt and dolerite, and partly disaggregated basaltic dykes, lie in a matrix of granophyre. Barkeval, the dark hill (*right*), is built of feldspathic peridotites.

been used to explain the relatively high gravity field over southern Rhum, a feature which the island shares with the Tertiary volcanic sites at the Cuillins of Skye and central Mull. The pronounced gravity 'high' over Rhum is attributable to the presence of a large mass of dense ultrabasic or gabbroic rock, probably in the form of an inverted steepsided cone which extends many kilometres into the Earth's crust; this represents the 'root' of the volcano.

The contacts between ultrabasic rocks and gabbros on one hand, and microgranites and felsites on the other are frequently very complicated. Blocks of dark coloured basalt and gabbro, and occasionally ultrabasic rocks, are set in light coloured matrices which are continuous with the adjoining felsites or granites. Good sections through the contacts are visible on the southern sides of Meall Breac and Cnapan Breaca, close to the east side of Minishal, and at either end of Harris Bay. The complex relationships are especially well displayed in the low cliffs on the headland at the east end of

Harris Bay, about 300 m south of Abhainn Raingail (figure
2.4). Here, granite veins and dykes intrude and brecciate the
basaltic and gabbroic rocks, giving the impression that the
granitic rocks are much the younger. This directly con-
tradicts the evidence from the volcanic centre as a whole: the
large-scale relationships deduced from careful mapping show
that the gabbros and ultrabasic rocks are the youngest intrusive rocks in the volcano. The explanation of this apparent
anomaly lies in the contrasted temperatures at which the
different rocks solidify and melt. Basalt magma, which
formed the gabbros and ultrabasic rocks, crystallises at about
1100°C whereas granitic magma, responsible for the felsites
and microgranites, crystallises at about 850°C. On Rhum,
hot basaltic magma has come against cold, solid microgranite
and felsite which it locally melted. At the same time, the
basalt magma was rapidly cooled by the earlier rocks and
formed a skin of solid basalt and fine-grained gabbro, which
probably developed contraction cracks as it cooled. These
cracks were invaded and exploited by the small amount of
granitic liquid present along the contact, thus forming the
complicated relationships depicted in figure 2.4. Similar features occur in many of the Tertiary volcanic centres; those
on Rhum are amongst the best in the region.

The Waning Stages of the Volcano

Only a small amount of igneous activity occurred on Rhum
after formation of the ultrabasic rocks. This third, final stage
was a time of rapid erosion when the top was stripped off the
volcano, exposing its roots down to a level not far short of
that seen today. A record of this stage is beautifully preserved
on the small, rugged hills of Fionchra, Orval and Bloodstone
Hill in north-western Rhum (figure 2.1). Thirty years ago,
G. P. Black demonstrated that on the northern slopes of
Orval the Tertiary basaltic lavas rested on a weathered,
eroded surface of Tertiary granophyre – the first time that it
had been shown that some of the Tertiary lavas of northwest
Scotland were younger than a volcanic centre.

The early Tertiary unroofing of the Rhum volcano can
also be convincingly demonstrated by examination of the
boulders and pebbles present in the conglomerates underlying the lava flows, which include pieces of microgranite,
felsite, gabbro and allivalite, all of which match rocks now

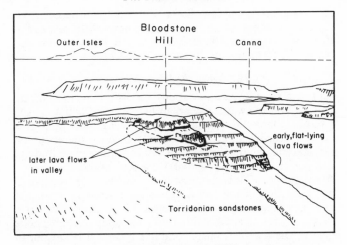

Figure 2.5. The east face of Bloodstone Hill showing two lava flows infilling a valley eroded in older lava flows and Torridonian sandstones. Based on a photograph taken looking north from Orval. (Sketch modified from figure 9 of the *Field Guide to the Tertiary Igneous Rocks of Rhum*, published by the Nature Conservancy Council.)

visible in the volcanic centre. In addition, there are many fragments of Torridonian sandstones and Lewisian gneisses. Water-lain sediments containing these deposits lie beneath basalt lavas at Maternity Hollow, north-east of Fionchra. The sediments also contain small fragments of carbonised logs and twigs, and similar deposits at about 350 m height on the northern slopes of Fionchra contain numerous delicate leaf impressions.

The clearest evidence that these sediments and lava flows filled valleys, and that valley formation, sediment accumulation and lava eruption all overlapped and recurred, comes from the east face of Bloodstone Hill where a steep-sided valley eroded in Torridonian sediments and Tertiary lavas has been filled by two later lava flows (figure 2.5). The sources of these and the other lava flows were not on Rhum and they may well have originated from a source east of Canna, where there is evidence for a small volcano.

Igneous activity on Rhum did not cease entirely after

accumulation of the lava flows. Further fissuring occurred since a few dykes cut through the lava flows at the east end of Fionchra. However, the majority of the dykes were intruded before the formation of the ultrabasic rocks.

The Age of the Rhum Volcano

Radiometric age determinations on Rhum rocks indicate that the volcano was active about fifty-nine million years ago – but how long did the volcanic activity last on Rhum? It is not yet possible to give a clear answer since the errors on radiometric age determinations are of similar magnitude to the period of activity – that is, a few millions of years. Another approach is to measure the ancient (or 'remnant') magnetisation of the rocks. When an igneous rock cools it becomes magnetised and a measurable component of this magnetisation is attributable to the contemporary magnetic field of the Earth. A fairly complete record of behaviour of the Earth's magnetic field is known from the present time to the Tertiary and earlier. The record contains frequent changes in the magnetic field; at times polarity was normal with respect to the present day, at other times it was reversed. The fifty-nine-million-year age of the Rhum volcano is close to a well-defined period of normal polarity. However, when the magnetisation of the Rhum igneous rocks was measured it was found that virtually all had reversed magnetisation. Thus, it is probable that all the intrusive igneous activity within the Rhum volcano, together with the subsequent erosion and effusion of lavas, took place between two periods of normal polarity. This limits the activity to a maximum of about three million years and it is possible that it all took place within the space of as little as one million years.

Post Tertiary Events

Probably the youngest event in the 'solid' geology of Rhum involved movement on the Long Loch Fault (figure 2.1). There appears to have been horizontal movement of about 500 to 800 m along the fault, in a right-hand sense when viewed across the fault. The relative direction of movement is clearly seen in the offset of an area of gabbro in Glen Harris (figure 2.1).

During the Pleistocene glaciations, the island was overlain by ice from the mainland. After the decay of this ice

sheet, local glaciers remained. The consequence of later
glaciation may be seen in the enlarged, steep-sided valleys of
Dibidil and Glen Harris, and in their numerous morainic
deposits. On Rhum, eleven local glaciers have been de-
lineated, the majority in deep valleys and steep corries of
Glen Harris, Fiachanis and Dibidil. Adjustments in sea level
allowed formation of the spectacular storm beaches at Harris
and Kilmory.

Soils on Rhum

There is a strong correlation between the soils on Rhum and
the solid geology. Glacial deposits are locally thick, and
important, but are generally of local origins. The soils on the
large areas of ultrabasic rocks and gabbros divide into shallow brown loams formed directly over rock, and deeper soils
in pockets and on colluvial slopes. The chemistry of the
ultrabasic rocks strongly influences the soils whose high
acidity is probably due to high exchangeable magnesium
relative to calcium, derived from weathering of the olivine-
rich rocks. The present soil conditions are often relatively
unstable compared with past conditions. Signs of former
stability are visible in several places, including Coire nan
Grunnd where well-developed podzol profiles occur beneath
colluvial cover.

Over the north and east of Rhum, the Torridonian
sandstone is often covered by a thin layer of peat, while
thicker deposits of peat floor much of Kinloch Glen and the
upper part of Kilmory Glen. An extensive tract of alluvial
sand is found in the lower part of Kilmory Glen and small
areas of marine sand dunes occur at Kilmory and several
points on the north coast, including Samhnan Insir. Wind-
blown sands also occur inland, south of Loch Coire nan
Grunnd and also at the Bealach an Oir, west of Askival.

The grassy basaltic lava hills of north-western Rhum
contrast strongly with the vegetation on peaty soils elsewhere. Unstable colluvial soils have developed on the steep
hillsides and these are sometimes subject to severe gullying
and erosion, as on the northern slopes of Fionchra. The
granitic rocks of Ard Nev and western Rhum tend to become
mantled by rock debris which can develop into screes and,
associated with this debris, there may be a development of
peaty podzols. On Monadh Dubh, north of Glen Shellesder,

weathering of the basal layer of sandy limestones in the Triassic rocks has contributed to the local, and for Rhum unusual, development of thin calcerous soils.

Further Reading

Bailey, E. B. (1945) Tertiary igneous tectonics of Rhum (Inner Hebrides), *Q. J. geol. Soc. Lond. 100*, 165-88.

Emeleus, C. H. (1983) Tertiary igneous activity, in *Geology of Scotland* (ed. G. Y. Craig) pp.357-97. Edinburgh, Scottish Academic Press.

Emeleus, C. H. (1980) *1:20,000 Solid Geology Map of Rhum*. Scotland, Nature Conservancy Council.

Emeleus, C. H. & Forster, R. M. (1979) *Field Guide to the Tertiary Igneous Rocks of Rhum*. London, Nature Conservancy Council.

Institute of Geological Sciences (1971) *Geological Survey of Great Britain (Scotland): One-inch sheet 60 (RHUM)*. Second provisional (Solid and Drift) edition.

Wager, L. R. & Brown, G. M. (1968) *Layered Igneous Rocks*. Edinburgh, Oliver and Boyd.

3

RHUM'S HUMAN HISTORY

J. A. LOVE

Prehistory

Compared with many other Hebridean islands, prehistoric remains on Rhum are few. In consequence, the island did not excite much interest from early antiquarians. Nonetheless, the history of man's presence is a long one and microliths found in recent excavations suggest that occupation dates back to the Mesolithic period. A cave at Bagh na h-Uamha may have been an early settlement site, with its charcoal hearth and large 'kitchen midden' of shellfish and animal bones; it is probably later than Mesolithic, but has yet to be satisfactorily dated.

Neolithic man penetrated the island's scrub forest to hunt deer, and an arrowhead of white siliceous stone was found recently on the slopes of Hallival. Two other arrowheads, both of pale green bloodstone, have been discovered on the coast – at Samhnan Insir in 1967 and at Kinloch in 1983. Rhum is the main source of this scarce mineral. It is often dark green with the tiny red spots from which it derives its name, but there is considerable variation in colour. Bloodstone's fracturing properties render it a convenient substitute for flint. Fragments and pebbles can be picked up from the shore at Guirdil below Creag nan Stearnain (Bloodstone Hill) and from here Neolithic man collected them to fashion into arrowheads and other implements. Several scatter-sites have been found; copious chippings have been uncovered in recent forestry ploughings to the north of Loch Scresort, for example, and in one of the farm fields at Kinloch – the site of a recent excavation.

Despite such human activity, traces of settlements on

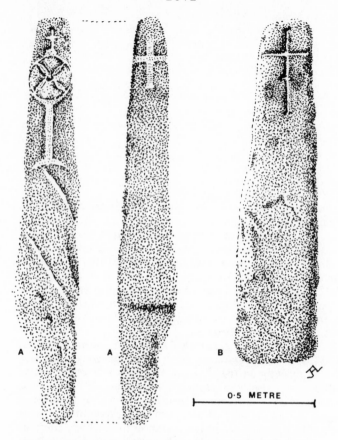

Figure 3.1. Celtic crosses on the Kilmory slab (*left* and *centre*) and on the pillar found on the beach at Bagh na h-Uamha (*right*).

Rhum are few and far between. A recent inventory by the Royal Commission on the Ancient and Historical Monuments of Scotland (RCAHMS) in 1983 includes medieval or later shielings, and it is possible that some of these may be on the sites of older settlements. It identifies several probable Bronze Age burial cairns and Iron Age promontory forts at Harris, Papadil and Shellesder.

In 1885 T.S. Muir found a sandstone pillar lying in the burial ground at Kilmory: it bore a plain Latin cross 25 cm tall. In 1925 the RCAHMS noted on its other face 'a small incised Latin cross surmounting a circular panel 8 inches in diameter . . . which contains a cross with expanding arms set saltire-wise' (figure 3.1). I cleared this of its obscuring encrustation of lichen a few years ago by turning it face down in the turf, which revealed more of its detail. Another, though less elaborate celtic cross slab was found lying face down in sand below high water mark at Bagh na h-Uamha: it has since been carried up the shore and set upright nearby. Both crosses probably belong to the seventh or eighth century AD and may have marked places of worship rather than graves. The name 'Kilmory' suggests there to have been a chapel in the vicinity, and its presence was noted (though not necessarily witnessed) by Martin Martin in 1704. Some authorities have identified this with a large ruin lying beside the burial ground but it is now agreed that this is a blackhouse.

The Irish Annals record the death in AD 677 of a St Beccan of Rhum, who may have lived as a solitary on the island, perhaps at the place which the Vikings later came to call 'Papadil' – the dale of the priests. The name of Rhum (which should be spelt 'Rum') may derive from another Norse word 'Romoy' meaning 'wide or spacious island'. Norse elements persist in many other placenames on the island, notably in the main hills and glens such as Askival, Trollval and Dibidil. There is scant evidence of Viking settlement on the island: a burial cist was found at Bagh na h-Uamha by Drew Smith in 1949 as well as an ivory gaming piece, 4.5 cm in diameter, which is decorated with a delicate interlace pattern. Nonetheless, for several centuries, the island was part of the Norse kingdom which stretched from Shetland to the Isle of Man.

Land Ownership

It was Somerled, a noble or petty king in Argyll of mixed Norse and Celtic blood who, in 1156, first wrested the southern Hebrides from the Norsemen. In 1266, following the defeat of King Haakon of Norway at Largs on the Firth of Clyde, the remainder of the Hebrides were ceded to the Scottish Crown. Somerled's descendants established on Islay a remote and powerful royal house which eventually came to be known as the Lordship of the Isles. Others of this noble descent – the MacRuaris, later to become Clanranald – gained estates on Skye, the Small Isles (including Rhum) and the adjoining mainland.

One of the many vassal clans of successive Lords of the Isles were the Macleans. Iain Garbh, a son of the seventh chief of the Macleans of Duart on Mull demanded the Isle of Coll as an inheritance from Alexander, the third Lord. This must have been around the middle of the fifteenth century. Iain, first Maclean of Coll, then added to his estates by extorting the Isle of Rhum from Clanranald: local tradition maintains that this was exchanged for a galley which subsequently proved unseaworthy. Whatever the bargain, Allan MacRuarie of Clanranald was reluctant to confirm the deal and was held prisoner on Coll for nine months. Thus Rhum came into the hands of the Macleans of Coll, a junior branch of Duart.

The Lordship of the Isles persisted as an independent realm until it was brought under control of the Crown by King James IV in 1493 when John, the fourth Lord, forfeited his title. Only when the vassal clan chieftains had submitted to the Crown did they have their estates returned to them with royal approval.

At last, Maclean of Coll could assert his independence from Duart, but the latter long resented this and repeatedly sacked the Coll estates. During this prolonged internecine feud, the High Dean of the Isles, Donald Munro, noted with appropriate clerical diplomacy how in 1549 Rhum 'pertained' to the Laird of Coll but 'obeys instantlie' to Maclean of Duart. Munro uses Coll's Gaelic patronym 'McAneAbrie' or Mac Iain Abrach while rendering the rival camp as 'McGillane of Doward'. The issue became further confused when a report submitted to King James VI later noted that Rhum was 'possest and in the handis of Clan-Ranald'. The latter at that time was known to be feuding with the Macleans so perhaps had seized an opportunity to retake the islands which his ancestor Allan MacRuarie had been forced to relinquish. Despite these temporary irregularities, Rhum effectively remained the property of the Macleans of Coll until it was sold to Lord Salisbury in 1845.

Population

The earliest documentary reference to the island's population is Fordun's *Chronicle*, written about the year 1400, but all it can offer is that Rhum had 'few inhabitants'. Nearly two centuries later, the report prepared for King James VI indicated that the island could muster only six or seven fighting men (between the ages of sixteen and sixty). Eigg, on the other hand, could raise sixty men so Rhum's population at that time must still have been small: it had certainly suffered repeatedly in devastating clan feuds. By 1728 the Society for the Propagation of Christian Knowledge recorded 152 persons over the age of five – a community perhaps equivalent to about 180 people.

A census of Scotland was undertaken in 1755 by Rev. Alexander Webster and another, more accurate one in 1764 by Rev. John Walker. The latter made a particular study of the Small Isles for which he obtained his statistics from the parish minister – 'a sensible and careful man'. In Scotland, official censuses commenced in 1801, four decades earlier than in England and Wales. Webster found 206 inhabitants on Rhum while Walker counted 304 (some versions of his data state 302). Such considerable increase in the island's population within only nine years may to some extent be the result of inaccuracies in Webster's enumeration. However, Walker added that there had been no recent outbreaks of smallpox and that the health of the islanders had improved greatly in the thirty years leading up to his census, due mainly to an increase in the amount of grain available and the introduction of garden produce, notably the potato. No doubt the suppression of clan warfare following the 1745 Rebellion too permitted a population increase.

In 1772 Thomas Pennant recorded 59 families on Rhum and a total of 325 inhabitants, while in his tour of 1785 John Knox estimated 300 islanders (though he may have been reiterating Walker's published figure). The Statistical Account of 1796 compiled by the parish minister on Eigg claimed that Rhum supported 443 inhabitants. This figure was repeated by Neckar de Saussure who visited the island in 1807 (not 1822 as is sometimes reported). By 1811 Macdonald put Rhum's population at around 350, a more realistic figure perhaps, which is verified by Alexander Hunter in

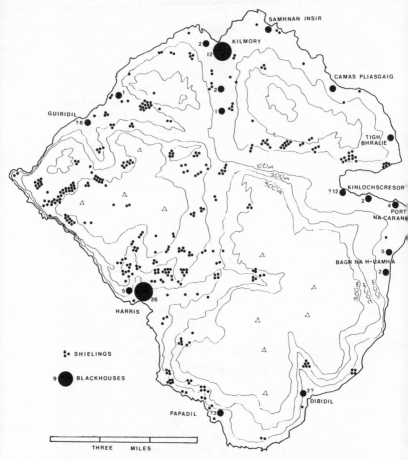

Figure 3.2. Map of Rhum showing the sites of permanent human settlement (the number of surviving blackhouse ruins is placed beside each large circle) and the distribution of shieling huts (small circles) in relation to altitude (contours at 100-m intervals).

1826. The latter cynically attested how it was common practice at the time to marry early – even at an age of only seventeen – so that couples could produce sufficient children to exempt the husband from military service. However, in

March 1793 eleven young men enlisted in the Breadalbane Fencible Regiment and further decreases in population at this time resulted from a number of small voluntary emigrations.

Agriculture and Settlement

In 1772 Thomas Pennant recorded nine villages on Rhum which, according to Langland's map of Argyll (1801), were located at 'Kilmory, Sandanisker, Camas Pleaseig, Kinloch Scresort, Cove (i.e. Bagh na h-Uamha), Glendibble, Pappadill, Harris and Guidle'. The surviving blackhouse ruins and lazybed cultivations bear this out (figure 3.2) although another pre-Clearance settlement occurred at Port na Caranean. A few other isolated ruins occur at remote situations such as Tigh Bhralie and Bagh Rudha Mhoil Ruaidh on the north-east coast. These houses tend to display 'late' features such as a tiny window, or even a fireplace at one end, so they were probably built as the population increased late in the eighteenth century. The sites may previously have been coastal summer shielings. A settlement and cultivations at Kilmory Fank (apparently once known as An Leth-pheighinn – 'the half penny', referring to a measure of land) is the only inland pre-Clearance settlement, all the others being coastal. Pennant mentioned only a dozen houses at Kinloch, although the foundations of only a few of them are still discernible nowadays (in what is now the campsite).

Harris was the largest single settlement comprising about thirty black-houses clustered together behind the raised beach from which they derived shelter from the prevailing winds (figure 3.3). A further twenty or so houses lie at the mouth of the Kilmory River with one or two others scattered along the shore to the east.

It is not easy however now to determine the number of actual houses on the island since some are intermediate in size between what were obviously houses (measuring from 6 m by 3 m to as much as 13 m by 4 m) and what could only have been barns or byres (most about 3 or 4 m but sometimes as little as 2 m square). Some of the intermediate-sized houses may have been relegated for use as byres or outhouses only when a larger and more substantial blackhouse came to be built alongside.

Around 100 or so blackhouse ruins are still discernible

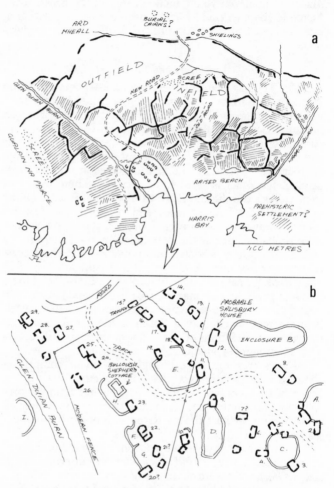

Figure 3.3. (a) Pre-Clearance runrig settlement at Harris, showing the fields of lazybeds each surrounded by stone and turf dykes. (b) Sketch map of Harris village, with about 30 blackhouse ruins, associated byres and enclosures. One house (no. 12) is of dressed stone and was probably built (or at least renovated) by Lord Salisbury. Another (not numbered) was built later by Bullough.

34

Figure 3.4. The Bullough mausoleum and shooting lodge at Harris, Isle of Rhum. The complex lazy-bed system is obvious in the shadows of the snow and in places can be seen to abut (and even continue over) the blackhouse ruins. It is suggested that this small collection of houses was abandoned at an early date, before the main Clearances in 1826-28.

on Rhum. These would have easily accommodated 400 people and agree with Walker's calculation of about five persons per household in 1764. Not all these houses would have been in use at the same time, some falling into disuse during voluntary emigrations prior to the main Clearance. One small group above the modern mausoleum at Harris on the slopes of Gualann na Pairc, are now partly obscured by lazybeds and must have been vacated at an early date (figure 3.4).

An extensive and well-preserved field system is still apparent at Harris and has been little altered by later agricultural developments. Fields of lazybeds were used to cultivate oats, perhaps some barley and, latterly, potatoes and kail. The meagre soil was heaped up into parallel ridges and manured with seaweed or soot-impregnated thatch, leaving drainage ditches between. The wet climate did not encourage

35

a)

b)

Figure 3.5. The three types of shieling found amongst the hills of Rhum. (a) One of the few complete shieling huts (although lacking its turf covering) on the island, a corbelled stone cell above Kinloch. (b) A good example of a turf-covered chamber (only the stone founds surviving) at Kilmory Glen, with a tiny door leading to a small corbelled stone cell for sleeping or the storage of dairy utensils. (c) A fine rectangular shieling (again minus its turf covering) near Papadil, a late design resembling a tiny cottage.

36

(c

many good harvests so that the island's economy tended more towards a pastoral one. Both cows and sheep, together with a few goats, were kept for milking; and some beef and mutton also became available from time to time. Ponies and oxen were kept as beasts of burden. While the menfolk tended the crops or repaired the cottages, the women and children took the stock to the hill grazing, where summer encampments or shielings had long been established.

The ruins of some 400 shielings are visible on Rhum (figure 3.2) and more are coming to light. All are within easy reach of rich grazings and they usually occur singly or in small groups of three or four huts, more rarely in groups of eight or nine. They are mostly located on the less steep hill slopes in the north around the 100-m contour; in the Harris area along a shelf some 200 m above the sea, and in Guirdil at about 300 m on a fertile plateau appropriately called Airigh na mathe-innis, 'the shieling of the good grazing'. Whenever possible, the huts were constructed close to a burn or stream with good grazing nearby and were often adjacent to scree slopes which provided abundant stones as building materials. Topographical features were prudently employed to afford shelter from the elements while the doors of many huts faced east, away from the prevailing winds.

There was considerable variety in the shape and structure of the shielings but three basic types might be recognised (figure 3.5). Twenty-four per cent of the huts on Rhum are rectangular, enclosing a space about 3 m by 2 m. These are the most recent design, as indicated by the mounds of collapsed turves and stones from earlier structures around and below them. The most frequent type (48 per cent) is a circular chamber 3 to 4 m in internal diameter (usually with a small stone cell attached) but it is also the most variable, some having additional chambers or several small cells attached. Although the walls of both chambered and rectangular huts may be only two or three courses high, a roof would have been constructed of turves over-lying a timber framework. The remaining huts (28 per cent of the island total) were large circular cells built almost entirely of and roofed with over-lapping stones: some of the most puzzling of these structures are linked by stone dykes. Such cellular huts seem to be the oldest of the three designs and bear a close resemblance to the corbelled cells of mediaeval anchorite monks, perhaps even exhibiting influences from more ancient prehistoric dwellings.

Rhum also possesses a fine example of a deer trap (figure 3.6) on the south slope of Orval in Glen Duian (grid ref. NM 329987). In 1549 Dean Munro referred to such structures as 'settis' or seats (sometimes erroneously interpreted as 'saitts'), and the *Old Statistical Account* of 1796 describes the mode of operation of a deer hunt or 'tainchell': 'on each side of a glen formed by two mountains, stone dykes were begun pretty high in the mountains and carried to the lower part of the valley, always drawing nearer till within three or four feet of each other. From this narrow pass, a circular space was enclosed by a stone wall of a height sufficient to confine the deer: to this place they were pursued and destroyed'. A contemporary account of a deer hunt on mainland Scotland mentioned how eventually 'with dogs, gunnes, arrows, durkes and daggers, all in the space of two hours, fourscore fat deare were slaine'. Red deer became extinct on Rhum in the 1780s.

The Clearances

During the first half of the nineteenth century, many parts of the Highlands were cleared of their inhabitants so that the land could be let more profitably to sheep graziers from the

Figure 3.6. Deer trap on the slopes of Orval (January 1980), looking downhill along one of the funneling stone dykes leading to a large circular enclosure whose walls are over 2 m high in places. The trap lies near the head of Glen Duian.

south. Rhum lost virtually all its inhabitants to the New World, but much confusion has prevailed regarding the circumstances and the exact date of this tragic event. Banks (1977), for example, claimed that it occurred in 1828, mainly because of a list which survives in official Nova Scotian archives, which names 208 emigrants – all said to have originated from Rhum – who arrived in Cape Breton aboard the ship *St Lawrence*. However one cannot dispute the evidence given to a Government Select Committee on Emigration by Alexander Hunter – an Edinburgh lawyer and the man who had actually supervised the Clearance for Maclean of Coll. He testified that 300 people were removed from Rhum in July 1826 and that 50 folk were allowed to remain. This emigration took place aboard two ships, the *Highland Laddie* and *Dove of Harmony*, which berthed in Nova Scotia that year. Apparently, the remainder of the Rhum folk followed in 1828, in the company of 150 islanders from Muck – another of Maclean of Coll's properties. Together they make up the 200 names on the *St Lawrence*'s passenger list. The *New Statistical Account* of 1845 confirms that only one family was allowed to remain on Rhum.

It has been a common misconception that the people had left Rhum entirely voluntarily. Certainly, Alexander Maclean of Coll was of a benevolent disposition, distributing charity to his beleaguered tenants and ultimately waiving their considerable rent arrears. It is likely to have been his more mercenary son, Hugh, who was running the estates at the time and who, in 1825, issued the islanders with a year's notice to quit. In his evidence to the Select Committee in 1827 Alexander Hunter confessed that although some of the Rhum folk accepted their fate, 'others were not willing to leave the land of their ancestors'. Edwin Waugh in his book *The Limping Pilgrim* (1882) described how a shepherd who had witnessed the event would never forget it to his dying day – 'the wild outcries of the men and the heart-breaking wails of the women and children filled all the air between the mountain shores'.

The islanders' passage to the New World was unlikely to have been a comfortable one. For the 37-day passage, Coll had provided each adult with 35 pints of water, 11 pounds of oatmeal, 3½ pounds of bread or biscuit, 1½ pounds of beef, half a pound of molasses, half a pound of pease or barley and a quarter of a pound of butter. (Hunter admitted that by replacing beef with so much oatmeal the laird's expenses had been considerably reduced.) A few islanders were able to eke out these rations with some salt mutton and potatoes of their own. Disease was probably rife. Official records in Nova Scotia revealed how one of the ships which had earlier been used in the Rhum Clearance berthed again the following year with twenty-two out of two hundred or more emigrants on board suffering from measles; eighteen others died during the voyage. The cause of such dire fatality – which was by no means unusual in emigrant ships – was 'readily traced to the confined, crowded and filthy state of the vessel'.

The one family remaining behind on Rhum were Mac-

leans who could claim direct descent through eleven generations from the Macleans of Coll themselves. Not surprisingly, they were unable to shepherd the 8,000 sheep brought
in by the new grazing tenant Dr Lachlan Maclean of Gal-
lanach. Therefore, another dozen families being evicted from
Bracadale on Skye and from Mull had to be imported as
labour. By 1831 Rhum supported a community of some 134
souls. However, within a few years, the whole sheep-farming
venture failed, bringing ruin to Dr Maclean and forcing his
landlord, Hugh Maclean of Coll, to sell the island in 1845.

Rhum as a Sporting Estate

The new owner was Lord Salisbury, the father of the Victo-
rian Prime Minister. He retained 5,000 sheep with attendant
shepherds and also developed the island's sporting potential.
Red deer were reintroduced and unsuccessful attempts made
to improve the game fishing – including the construction of a
dam at the head of the Kilmory river which burst two days
after its completion. Such projects provided welcome
employment on Rhum and its neighbouring islands at a time
when a potato famine was rife throughout the Hebrides.
Salisbury was intolerant of unemployed paupers and in 1852
he 'encouraged' several more families on Rhum to emigrate
to Canada.

In 1869 the island again came on the market and was
sold to a family of Campbells of Ballinaby. Sheep and deer
interests were maintained although no improvements were
made until Rhum passed into the hands of a wealthy Lancashire industrialist, John Bullough, in 1886. Comfortable
shooting lodges were erected at Harris and Papadil, and, in
1889, 80,000 trees were planted around Loch Scresort. John
Bullough died in 1891 and within ten years his son and heir,
George, had built an extravagant and impressive 'castle' of
red sandstone, hewn on the Isle of Arran. It incorporated
numerous innovative features including hydro-electricity
and air-conditioning. The oak-panelled interior with its
plush Edwardian furnishings included a bizarre assortment
of Oriental bric-a-brac collected on George's recent world
cruise. Top-soil imported from Ayrshire contributed to the
extensive lawns and gardens in the Castle policies, some 200
m of the walled garden being under glass, a turtle pond being
maintained at the mouth of the Kinloch River. The Bul-

loughs remained in residence for only a few months each year and, after the 1914–18 war, the Castle and its grounds had begun to assume a slight air of decay. In 1957, Sir George's widow (who died twelve years later at the age of 98) sold the island as a nature reserve to the Nature Conservancy Council.

Further Reading

Banks, N. (1977) *Six Inner Hebrides*. David & Charles, Newton Abbot.

Love, J. A. (1980) Deer traps on Rhum, *Deer 5*, 131-32.

Love, J. A. (1981) Shielings of the Isle of Rhum, *Scott. Studies 26*, 39-63.

Love, J. A. (1983) *The Isle of Rhum – a short history*.

Royal Commission on the Ancient and Historical Monuments of Scotland (1983) *The Archaeological Sites and Monuments of Scotland No. 20. Rhum (Highland Region)*. Edinburgh, HMSO.

Ritchie, P. R. (1967) The stone implement trade in third-millennium Scotland, in *Studies in Ancient Europe* (eds J. M. Coles and D. D. A. Simpson) pp. 117-36. Leicester University Press.

4

BOTANY, WOODLAND AND FORESTRY

M. E. BALL

Post Glacial Forest

About 7,000 years ago the sheltered low ground in the glens and eastern seaboard of Rhum below the 300-m contour was well wooded (figure 4.1). Analysis of plant remains and pollen in a bog near the Long Loch, at an altitude of 160 m, shows that woodland developed 13,000–11,000 years ago. Subsequently, there was partial reinvasion of the ice sheet (the Loch Lomond Readvance), when woodland of willow *Salix* sp., birch *Betula* sp., pine *Pinus sylvestris*, bog myrtle *Myrica gale* and/or hazel *Corylus avellana* (with indistinguishable pollen) and juniper *Juniperus communis* established in the prevailing tundra vegetation of sedges, club moss and heath. Tree and shrub growth declined during the following cold phase but, subsequently, with rising temperatures less than 10,000 years ago, a long period of woodland succession and development proceeded. Birch, pine and particularly bog myrtle (possibly with hazel) increased markedly, later to be followed by the more warmth-loving ash *Fraxinus excelsior*, alder *Alnus glutinosa*, elm *Ulmus glabra* and oak *Quercus* (spp), with a decline in pine. The maximum extent of this primaeval forest is shown in figure 4.1.

Forest Clearance

The warmer period coincided with the activities of man, and a sudden upsurge in bracken *Pteridium aquilinum*, juniper and herbaceous plants, with comparatively rapid fluctuations in tree pollen suggesting the first tree-felling and woodland clearance. Then followed a recovery of pine and birch, rises in alder, heathers and water plants and a decline in elm and oak associated with an increasingly cooler and wetter climate. Though this probably led to some bog extension into the forest, neither the activities of man nor climatic deterioration caused significant forest destruction until about 2,500 years ago. Later Norse intrusions in the eighth century with their burning, grazing and cultivation finally brought about a predominantly treeless landscape.

The Last Remnants

Small fragments of natural scrub still survive in steep gullies and cliff ledges of Rhum where they are inaccessible to grazing and protected from fires. These comprise mixtures of hazel, downy birch *Betula pubescens*, oak, rowan, holly, aspen, hawthorn, grey sallow *Salix atrocinerea* and eared sallow *S. aurita*. Alder, ash, elm and Scots pine are extinct in these remnants, although one old ash remains at Kinloch, but they were all reintroduced within the past century at Kinloch and Papadil. The best examples botanically are found in the stream gorges of the east and south coasts, and especially in glens Shellesder, Kinloch, Kilmory, Camas Pliascaig and Caves Bay, and they show significant survival of a woodland field layer and an epiphytic lichen flora. The woodland field layer comprises wood sorrel *Oxalis acetosella*, ivy *Hedera helix*, angelica *Angelica sylvestris*, yellow pimpernel *Lysimachia nemorum*, primrose *Primula vulgaris*, herb Robert *Geranium robertianum*, bluebell *Hyacinthoides nonscriptus*, wood violet *Viola riviniana*, stitchwort *Stellaria holostea*, bitter vetch *Vicia orobus*, wood rush *Luzula sylvatica*, mountain fern *Oreopteris limbosperma*, golden scaled fern *Dryopteris pseudomas*, lady fern *Athyrium filix-femina* and royal fern *Osmunda regalis*. Several more woodland plants survive in shady places where no trees remain, among which are wild garlic *Allium ursinum*, valerian *Valeriana officinalis*, red campion *Silene dioica*, dog's mercury *Mercurialis perennis*, cow wheat *Melampyrum pratense*, celandine *Ranunculus ficaria*, tutsan *Hypericum androsaemum*, water avens *Geum rivale*, sanicle *Sanicula europaea*, bugle *Ajuga reptans*, enchanter's nightshade *Circaea lutetiana*, slender false brome *Brachypodium sylvaticum* and hay scented buckler fern *Dryopteris aemula*. The only significant woodland plants absent from the relict forest flora of Rhum, which occur quite commonly in neighbouring island woods, are woodruff

Figure 4.1. Probable distribution of woodland before deforestation.

Galium odoratum, wood avens *Geum urbanum,* melancholy thistle *Cirsium heterophyllum* and the woodland grasses melick *Melica uniflora* and *M. nutans* and wood brome *Zerna ramosa.*

Although there was still some woodland recorded in

45

1703, the final extinction of the last copse of wood was recorded by McLean in 1796. After this, the island was virtually treeless for half a century until a half-acre plantation of sycamore, beech, elm and ash was established round Kinloch House in the middle of the nineteenth century. The Kinloch policy woods which exist today were planted about eighty years ago by the Bullough family. Of the many native and exotic trees planted at this time the most prominent today are Scots and Corsican pines, larch, Norway spruce, lime, sycamore, Norway maple and sweet and horse chestnut in addition to the native broadleaved trees mentioned above, together with alder and birch. Altogether 120 tree species have been catalogued and, though the woods do not resemble the natural forest, they have enabled many of the native woodland plants to survive, including attractive carpets of bluebell, wood sorrel and buckler fern, and many species of epiphytic lichen and moss which depend on shade and high humidity. Some more local woodland plants, including wood anemone *Anemone nemorosa* and wood sedge *Carex sylvatica*, are now extinct outside the Kinloch woods.

Botanical Studies

On 11 March 1886, the botanist Symington Grieve read a paper of considerable length to the Botanical Society of Edinburgh after nine days' botanising on the Isle of Rhum in July 1884. In this first floristic account of the island, he recorded 198 flowering plants including such rarities as long-stalked pondweed *Potamogeton praelongus*, wood bitter vetch *Vicia orobus*, Scottish asphodel *Tofieldia pusilla* and long-stalked yellow sedge *Carex lepidocarpa*. While most of the plants on the list are still found, a few have been rejected as unlikely on distributional grounds and others have not been recorded since and are probably extinct. These include harebell *Campanula rotundifolia* (rediscovered at Harris in 1985), spring squill *Scilla verna*, alpine cinquefoil *Potentilla crantzii*, hairy buttercup *Ranunculus sarduous* and melancholy thistle *Cirsium heterophyllum*. Symington Grieve also studied the mosses and liverworts in 1884, reporting his findings in 1889 to the Botanical Society of Edinburgh.

In 1939 and 1941 W. A. Clark and J. W. Heslop-Harrison published their thorough botanical survey of the Hebrides including Rhum. Many more higher plants were

Figure 4.2. The rare Alpine penny cress *Thlaspi alpestre*, first found by Professor Heslop-Harrison on Fionchra.

discovered, among which were the rare alpines Norwegian sandwort *Arenaria norvegica* and alpine penny cress *Thlaspi alpestre* (figure 4.2). However, some of Heslop-Harrison's most unusual plant records may have been mischievously introduced. Among the eleven doubtful records which have been rejected on distributional grounds are alpine catchfly *Lychnis alpina*, rock whitlow grass *Draba norvegica*, alpine pearlwort *Sagina normaniana,* and six sedges including two new to Britain, *Carex bicolor* and *C. glacialis.* Whatever the truth of the matter, Heslop-Harrison returned to Rhum in 1961 in his old age and apparently re-found *Carex bicolor.* However, it has not been seen since that visit.

After the Nature Conservancy purchased the island in 1957, comprehensive lists of all Orders were compiled and included 48 lichens, 146 mosses, 55 liverworts and 289 flowering plants and ferns. In 1965, a check list for the island was published in the Transactions of the Botanical Society of Edinburgh by W. J. Eggeling. The lichens remained under-recorded until 1981 when O. L. Gilbert visited Rhum in two successive seasons, and as a result of these surveys added about a hundred species to bring the total number to 352, including a species new to Scotland, *Lecanactis plocina.*

Since the 1965 check list, a further twenty-four plants have been recorded among which are sea rush *Juncus maritimus* in 1967, dog's mercury *Mercurialis perennis* in 1971, white campion *Melandrium album* in 1972, oyster plant *Mertensia maritima* in 1974, barren strawberry *Potentilla sterilis* and wood sedge *Carex sylvatica* in 1976, chestnut sedge *Blysmus rufus* in 1981 and corn salad *Valerianella locusta* in 1983, bringing the total of higher plants and ferns to 590 taxa.

Plant Ecology

To provide baseline information on the vegetation in order to assess trends and successions resulting from long-term management, Nature Conservancy botanists established a series of thirteen experimental plots on grasslands and heaths. In addition, the distribution of the nine major plant communities (see table 4.1) was mapped by R. E. C. Ferreira.

Moorland, including wet heath, fen and blanket bog, covers Rhum's wet eastern hills and glens on the peaty soils overlying Torridonian sandstone rock. The more floristically rich and fertile herb-rich heaths and *Agrostis-Festuca* grasslands occupy much smaller areas but are widely distributed in small pockets on the lower slopes of the western glens and coastal cliff terraces. *Calluna* and *Nardus* heaths on acid mineral soils, and marshes in wet hollows are also confined to the south-western hill slopes and coastal cliff-ledge communities and scrub woodland relics are the only true natural vegetation. When the Nature Conservancy took over the

Table 4.1. Nine principal plant communities found on Rhum.

Formation (mapped at 1:20,000)	Associations (plotted at 1:12,000)
1. Calluna heath	*Calluna* heath
2. Wet heath	a) *Calluna-Trichophorum-Molinia* b) *Rhacomitrum-Calluna*
3. Blanket bog	a) *Eriophorum-Calluna* b) *Eriophorum* c) *Luzula sylvatica*
4. Nardus heath	a) *Nardus* b) *Nardus-Calluna* c) *Nardus-Juncus squarrosus*
5. Schoenus fen	a) *Schoenus-Molinia* b) *Schoenus* (bog) c) *Schoenus* (flush)
6. Molinia flush	a) *Molinia* (flush) b) *Molinia* (grassland)
7. Herb-rich heath	a) *Calluna* (calcium-rich soils) b) *Calluna* (magnesium-rich soils) c) *Vaccinium-Calluna*
8. Marsh	a) *Juncus acutiflorus*
9. Agrostis-Festuca grassland	a) *Agrostis-Festuca* (species-rich) b) *Agrostis-Festuca* (species-poor) c) *Rhacomitrium-Festuca-Vaccinium*

management of Rhum in 1957, its hill pastures were grossly overgrazed and the vegetation was in need of a period of recovery with reduced grazing impact. Some 2,000 sheep and 40 cows were removed in 1957, leaving the island grazed only by 1,600 red deer and locally by feral goats and ponies. The effects of this change in land use varied with each plant community. Whereas on wet heath vegetation, reduced grazing had virtually no effect, on herb-rich heaths and grasslands the number of species fell by about one-third with progressive loss of species characteristic of well grazed swards such as wild thyme *Thymus drucei*, purging flax *Linum catharticum*, eyebrights *Euphrasia* spp, plantains *Plantago* spp, self heal *Prunella vulgaris*, birdsfoot trefoil *Lotus corniculatus*, clovers *Trifolium* spp, daisy *Bellis perennis* and yarrow *Achillea millefolium*. There were corresponding increases in the taller growing or tufted species like fescues *Festuca* spp, purple moor grass *Molinia caerulea* and bell heather *Erica cinerea*.

To reverse these changes, a herd of free-ranging highland cattle was established in the early 1970s and subsequently they have built up to some 60 beasts, hefted to the Harris area in the south west for breeding and wintering, and summering in Guirdil and Shellesder in the north west. Many of the plants which declined during the previous period are now regaining their former abundance, though some species, including ribwort plantain *Plantago lanceolata*, heath bedstraw *Galium saxatile* and field buttercup *Ranunculus acris* have failed to recover their former abundance, possibly because they are more suited to the old crofting land-use with periodic cultivation or sheep grazing.

The effect of completely excluding grazing within fenced enclosures has been studied using fenced quadrats, and shows similar trends. The number of plant species surviving after 5–20 years of grazing exclusion are reduced by more than half, while taller growing plants spread and take their place. On ungrazed moorland vegetation, heather and purple moor-grass respond the most, whilst in grasslands the vegetation becomes a thick mat of tall growing and aggressive grasses like red fescue *Festuca rubra*, viviparous fescue *F. vivipara*, meadow grass *Poa pratensis* and creeping soft grass or yorkshire fog *Holcus* spp, with the later invasion by false oat grass *Arrhenatherum* and umbellifers such as wild angelica *Angelica sylvestris*. Due to the lack of available seed sources woodland regeneration is insignificant except by recovery of shrubs like holly and hazel hitherto suppressed by browsing.

Plant Distribution

The variety of plants of maritime grasslands, heaths and marshes is best seen in Kilmory glen where a sandy strand and small coastal marram dune system gives way to a machair plain of dry calcareous grassland rich in wild carrot *Daucus carota*, kidney vetch *Anthyllis vulneraria*, centaury *Centaurium erythraea*, lesser meadow rue *Thalictrum minus* and field gentian *Gentianella campestris*. The herb-rich heath knolls have birdsfoot trefoil, wild thyme, purging flax, catsear *Hypochoeris radicata* and self heal, while the damp

Figure 4.3. Grass of Parnassus found in marshy habitats, particularly at Kilmory.

meadows have grass of parnassus *Parnassia palustris* (figure 4.3), sneezewort *Achillea ptarmica*, moonwort *Ophioglossum vulgatum*, adder's tongue *Botrychium lunaria*, red rattle and lousewort *Pedicularis palustris* and *P. sylvatica*. Frog orchid *Coeloglossum viride* (figure 4.4) and a range of marsh and meadow orchids also occur, together with a rarity special to Rhum, the horsetail *Equisetum* × trachyodon. The grasslands and herb-rich heaths of Harris glen, draining from the ultrabasic rocks, are different in character and less species-rich. Like Kilmory, they have a wide variety of meadow grasses which include the northern atlantic species viviparous fescue, crested hair grass *Koeleria gracilis* and heath grass *Seiglingia decumbens*, and a particularly characteristic flower is the mountain everlasting *Antennaria dioica* (figure 4.5) which occurs almost from sea level to the tops of the hills. The fertile Glen Guirdil, draining the basalt hills Fionchra and Bloodstone Hill, also has its own character and its grasslands have some woodland relict flora like cow wheat *Melampynum sylvaticum*, primrose, bilberry *Vaccinium myrtillus* and woodrush *Luzula sylvatica*. The Rhum grasslands are in addition of special interest for the variety of eyebrights *Euphrasia* spp.

Wet heaths and blanket bogs cover a large proportion of the total land surface, and are characterised by the predominance of *Molinia* and the virtual absence (except for a few plants at Kinloch) of bog myrtle *Myrica gale*, so common in the rest of the Hebrides. Past heavy grazing and burning has caused a widespread reduction in the bog mosses, although there are still sixteen species recorded, and *Sphagnum magellanicum*, *S. rubellum* and *S. subsecundum* survive in a few places. The insectivorous sundews *Drosera angelica*, *D. intermedia*, *D. rotundifolia* and *D.* × *obovata*, butterworts *Pinguicula vulgaris*, *P. lusitanica* and bladderworts *Ultricularia minor*, *U. neglecta* and *U. intermedia* are all well represented. Particularly abundant throughout wet moorlands are bog asphodel *Narthecium ossifragum* and heath spotted orchid *Dactylorrhiza maculata*. An interesting feature of blanket bogs and flushes draining from ultrabasic rocks is the abundance of black bog rush *Schoenus nigricans* which over large areas is often co-dominant with *Molinia*. The rare brown-beaked sedge *Rhynchospora fusca* occurs in a small raised bog at the head of Kinloch glen.

The sea cliffs and cliff terraces, while often spectacular scenically, are unremarkable botanically, not differing significantly from other Hebridean islands. Commonly rep-

Figure 4.4. Frog orchid, found on short grass swards at Kilmory.

resented are lovage *Ligusticum scoticum,* roseroot *Sedum rosea* and tall herb flora such as water avens *Geum rivale* and wild angelica on ledges. Occurring rarely in these habitats are thyme broomrape *Orobanche alba,* pyramidal bugle *Ajuga pyramidalis* and early purple orchid *Orchis mascula*. During his vegetation survey R.E.C. Ferreira found the rare fern, forked spleenwort *Asplenium septentrionale* on rocks at Papadil.

The mountain flora, which has been least affected by land-use, is probably of greatest importance for conservation. The high levels of nickel, chromium, cobalt and especially magnesium, coupled with a lack of phosphate and potassium in soils derived from the ultrabasic rocks of the Rhum mountains, has riven rise to an unusual vegetation with open plant communities and much bare ground. Only in association with the Manx shearwater colonies has a closed montane grassland sward developed in response to the fertilising effect of these seabirds. Throughout the higher terraces of Hallival, Askival, Trollval and Barkeval the vegetation has developed as an Upland *Agrostis-Festuca* turf, closely cropped by red deer. Herbs and grasses more characteristic of lowland grasslands contain species such as sweet vernal grass *Anthoxanthum odoratum*, dog violet *Viola riviniana*, heath bedstraw *Galium saxatile* and heath speedwell *Veronica officinalis*.

In contrast, more unstable and exposed surfaces outside the influence of the shearwater colonies, with much bare rock, stony frost-sorted terraces and skeletal soil profiles, have a sparse and open-herb rich-heath vegetation on the ultrabasic rocks. Heather, *Agrostis*, *Festuca*, purple moor grass, heath grass, wild thyme, tormentil and the hair moss *Rhacomitrium lanuginosum* are the most frequent plants, together with alpine plants which thrive on these soils: mountain everlasting, mountain sorrel *Oxyria digyna*, stone bramble *Rubus saxatilis*, alpine bistort, *Polygonum viviparum*, alpine ladies mantle *Alchemilla alpina*, stiff sedge *Carex bigelowii*, northern rock cress, *Cardaminopsis petraea*, purple saxifrage *Saxifraga oppositefolia* and moss campion *Silene acaulis*. Two other alpines on these open windswept peaks are specially characteristic of Rhum, the Scottish asphodel and mossy cyphel *Cherlaria sedoides*. The rare Norwegian sandwort is confined to the slopes of Ruinsival.

Mountain vegetation on the 'granite' summits and ridges of Ainshval, Sgurr nan Gillean in the south and the Orval and Ard Nev to Sgorr Reidh triangle of high ground in the west comprises more closed plant communities. Here a

Figure 4.5. Mountain everlasting, a characteristic flower at Harris and on the western hills.

low wind-pruned sward of tundra vegetation, well grazed by deer, comprises alpine moss heath and *Nardus* grassland with heather, hair moss, wavy hair grass *Deschampsia flexuosa*, bilberry and crowberry *Empetrum hermaphroditum*. Stiff sedge, mosses and lichens, and in the wetter hollows heath rush *Juncus squarrosus* and least willow *Salix herbacea* which covers extensive summit swards, are also important components of the vegetation mosaic. Other rare willows of former alpine scrub are tea-leaved willow *Salix phylicifolia*, dark-leaved willow *S. nigricans* and *S. myrsinites* though the latter has no recent record.

Fionchra, with its contrasting north- and south-facing cliffs of basaltic rocks, has attracted botanists from successive generations. The alpine plants of the moist north facing crags include starry, mossy, kidney and arctic saxifrage, *Saxifraga stellaris*, *S. hypnoides*, *S. hirsuta* and *S. nivalis*. Golden saxifrage *Chrysosplenium oppositefolium* and yellow mountain saxifrage *S. aizoides* occur in other locations too, while purple saxifrage is also widespread. The alpine *Cruciferae* are also well represented on Fionchra, and on Bloodstone Hill and Orval; these include thale cress *Arabidopsis thaliana*, hairy rock cress *Arabis hirsuta*, northern rock cress *Cardaminopsis petraea*, hoary whitlow grass *Draba incaria* and the rare alpine penny cress on the south slopes of Fionchra and summit cliff of Bloodstone Hill.

Limestone soils, which often yield the most floristically rich vegetation in upland habitats, are poorly represented and occur in two small patches on Rhum, a Triassic limestone ridge on Monadh Dubh in the north west, and a Jurassic limestone crag on the slopes near Allt na Ba in the south. The only upland calcicole is found in a single locality (on Monadh Dubh); it is mountain avens *Dryas octopetala* growing in a herb-rich heath community amongst heather and bell heather.

Despite the abundance of open water habitats, the many shallow lochans in the hills lack the biological richness of some other habitats, and most are acid and nutrient-poor. However, many have rafts of white water lily *Nymphaea alba*, bogbean *Menyanthes trifoliata*, bur-reed *Sparganium angustifolia*, water pepper *Polygonum hydropiper* or pondweeds *Potamogeton* spp. The twelve pondweeds of the aquatic flora of Rhum are of most interest and their distribu-

56

tion requires further study. The submerged aquatics shore-weed *Littorella uniflora*, water lobelia *Lobelia dortmanna*, pillwort *Pilularia globulifera*, quillwort *Isoetes lacustris*, and stoneworts *Nitella* and *Chara* are all recorded from various lochans, and the sparsely occurring emergent swamp vegetation of loch margins comprises either bottle sedge *Carex rostrata*, water horsetail *Equisetum fluviatile* or common reed *Phragmites communis*. Neither bulrush *Schoenoplectus* nor saw sedge *Cladium* are recorded.

Recovery of Woodland

Since 1958, a sustained programme to rebuild the forest ecosystem has been implemented by the Nature Conservancy. As native trees and shrubs were unobtainable, a tree nursery was established at Kinloch in 1960 for an annual planting programme and the largely forgotten techniques of raising native species, no longer applicable in the big commercial nurseries specialising in exotic conifers and ornamental hardwoods, were redeveloped. Some 50,000 plants a year of (twenty species) each requiring individual treatment for successful germination and growth are produced. Standard nursery techniques are used: broadcast sowing onto prepared seedbeds one metre wide on light loam soils of pH 4–5, or alternatively on 'Dunemann' beds of needle leaf litter (figure 4.6). Seed is covered with a light layer of fine acid-rock grit. One-year or exceptionally two-year seedlings are transplanted using lining-out boards spacing rows 25 cm apart. Pre-germination and inter-row chemical herbicide treatments as well as hand weeding are found necessary.

Some native species required specialised treatment. Birch and alder with their small seeds vulnerable to heavy losses following germination were found to yield satisfactory seedling production only under an overhead irrigation system, notwithstanding Rhum's notoriously wet climate. The seed of berried shrubs, including holly, hawthorn, rowan, bird cherry and blackthorn, requires pre-germination treatment by up to three years' stratification in sand, or by freezing. Sallows, with their brief seed life, are raised by laying whole branches bearing mature catkins on the seedbed to allow immediate germination following natural seed fall. Alternatively, selected clones are raised from 2 cm × 40 cm stem cuttings raised for one or two years in nursery soil.

Figure 4.6. Making up Dunemann seed-beds
with conifer leaf litter.

Using the botanical surveys, native trees were planted
out in their preferred habitat. Thus the more demanding
ash, hawthorn, elm, aspen, bird cherry and hazel were confined to well-drained base-rich soils covered by *Agrostis-Festuca* grasslands; oak, birch and rowan to podsolised soils
dominated by grassy and herb-rich heaths; Scots pine with
birch and rowan to *Calluna* heath and wet heath on podsols
and peaty podsols, and alder, sallows, aspen and ash on fen
communities. Trees were planted in group mixtures to simulate the character of native woods.

Ploughed and fertilised wet heath and blanket bog vegetation is planted with the less demanding Scots pine, birch
and rowan, while unploughed purple moor grass *Molinia
caerulea* flushes support alder when treated with phosphate
fertiliser. With increasing altitude and exposure and on
impoverished peats characterised by deer sedge *Trichophorum caespitosum* near the natural tree line it is more
efficient to restore native woodland after establishing a
generation of the hardy exotic conifer lodgepole pine *Pinus
contorta*. At the 'thicket' stage, ground conditions are ameli-
orated sufficiently to enable native trees to be introduced into

Figure 4.7. Well-established oak and ash planted in a fertile but sea-exposed grassland site in the shelter of lodgepole pine and whin thickets at Harris.

cleared glades. On poor mineral soils the soil-improving pioneer shrubs whin *Ulex europaea* and broom *Sarothamnus scoparius*, whose root nodules fix atmospheric nitrogen, are planted.

The recovery of woodland on the saturated open moorlands is possible only in the absence of grazing and browsing animals. Even small numbers of deer destroy any seedling trees, whether they appear naturally or are planted, and successful re-establishment of the forest depends on effective deer-proof fences. These have to remain effective for at least thirty years until the new generation of trees is maturing and itself regenerating.

The Developing Woodland

Successful woodland establishment within the trial plots was followed in the mid-sixties by the planting of some 30,000 trees annually within a 600-ha enclosure extending from lower Kinloch glen and the north side of Loch Scresort to

Camas Pliascaig on the east coast. Birch, alder, Scots pine, rowan and grey sallow are the main trees planted on the predominantly acid moorland.

By 1970, the north side of Loch Scresort was fully planted, except for the rocky ridges, waterlogged mires, and a proportion of open moorland left purposefully as permanent glades or for later woodland succession. A 900-ha enclosure was then erected covering lower Kinloch glen and extending over the south side of Loch Scresort and Cave Bay. This area will be planted during the 1980s and a third phase will eventually extend to Kilmory and Glen Shellesder. More fertile sites will support stands of ash, elm and hazel in contrast to the pine–birch–oak stands of the acid Torridonian formation, which have supported the bulk of the plantations to date.

The development of the characteristic woodland field layer vegetation is slow and may require a programme of translocation of plants of low colonising ability, like dog's mercury and wood anemone. However, the more common woodland plants, especially those surviving as relict populations in heather and bracken on deforested ground, are already recovering naturally in some of the young plantations.

The first effects of afforestation are those of protection from grazing rather than woodland succession. Heather communities are frequently invaded by bracken, bent–fescue *Agrostis–Festuca* grasslands become dominated by meadow grass *Poa pratensis* or yorkshire fog *Holcus lanatus*, wet heaths by heather *Calluna vulgaris* and *Molinia* and rushy grasslands by *Molinia*. Later, other tall herbs including false oat *Arrhenatherum elatius*, tufted hair grass *Deschampsia caespitosa* and angelica gain prominence.

The shading and leaf-fall effects of afforestation on the field layer are more important. Canopy closure is associated with a reduction or disappearance of heather on former heaths, the replacement of fescue by yorkshire fog on grasslands, and an increasing dominance of purple moor grass on flushed ground, accompanied by a general reduction in the number of pastureland grasses and herbs.

As the slower growing trees like oak and ash are only beginning to close canopy, marked effects are limited to plantations of Scots pine, birch and alder. Heaths under pine and birch are beginning to show the first signs of succession

Figure 4.8. 25-year-old birch planted in 1958 in
a sheltered ravine in Kilmory Glen.

to a field layer consisting of wavy hair grass, feather moss,
bilberry and wood sorrel. Grasslands under birch and other
broadleaves are changing to a flora rich in woodland herbs of
highland birchwoods whilst under alder the field layer becomes a tall herb-fern mire (figures 4.7 and 4.8).

True woodland ground flora plants already recorded

under juvenile plantation canopies within the tree plots include honeysuckle, *Lonicera periclymenum,* hard fern *Blechnum spicant,* mountain fern, buckler fern, wavy hair grass, tufted hair grass, bluebell, angelica, wood sorrel, primrose, wood violet, self heal, pignut *Conopodium majus,* germander speedwell *Veronica chamaedrys* and bilberry.

To date well over half a million trees have been established covering an area of some 250 ha, and though growth rates on the poorer sites are relatively modest, substantial change in the natural history of the environment is apparent over wide areas. It is intended that those parts of the island which would naturally support forest and scrub habitats will eventually be reafforested.

Further Reading

Ball, M. E. (1974) Floristic changes on grasslands and heaths on the Isle of Rhum after a reduction or exclusion of grazing, *J. Envt. Mgt 2,* 299-318.

Ball, M. E. (1983) Native woodlands of the Inner Hebrides, *Proc. Roy. Soc. Edinb. 83B,* 319-39.

Ball, M. E. and Wormell, P. (1975) The nursery production of native Scottish trees and shrubs, *Scottish Forestry 29,* 102-10.

Eggeling, W. J. (1965) Check list of the plants of Rhum, Inner Hebrides. Part 1. Stoneworts, ferns and flowering plants, *Trans. Bot. Soc. Edinb. 40,* 60-9.

Ferreira, R. E. C. (1967) *Community Descriptions in Field Survey of Vegetation Map of the Isle of Rhum.* Unpublished typescript, Nature Conservancy.

Gilbert, O. L. (1982) *Lichen Survey of Rhum – Inner Hebrides.* Nature Conservancy Council.

Watling, R. (1969) Check list of the plants of Rhum, Inner Hebrides. Part 3. Fungi, *Trans. Bot. Soc. Edinb. 40,* 497-535.

Wormell, P. (1968) Establishing woodland on the Isle of Rhum, *Scottish Forestry 22,* (3), 207-20.

5

INVERTEBRATES OF RHUM

P. WORMELL

Introduction

In the process of cataloguing life forms which exist within a
Nature Reserve, the vast field of invertebrates, which includes worms, slugs and snails, millipedes and slaters, freshwater crustaceans, spiders and insects, tends to be neglected.
On Rhum invertebrates have been studied in greater detail
than in most Nature Reserves and yet there are still great
gaps in our knowledge and only when these gaps have been
filled can a true appraisal be made of the interrelationships
which exist between animals and plants. It is upon this basic
information that management policies can be formulated.

Invertebrates feed on plants or on other animals, either
dead or alive: there are even insects which feed on insects
which feed on insects. Many exist in water, others in dung.
There are many which subsist on decaying flesh and many
more on decaying fungi. There are invertebrates which live
in the nostrils of deer, in their livers, in their lungs, on and
under their skins, in their stomachs and intestines. Invertebrates seem to fill every possible niche.

A thorough knowledge of the distribution of these less
easily observed animals, many of which are very difficult to
identify, is essential in an outdoor laboratory such as Rhum.
Many of the birds recorded on Rhum feed almost exclusively
on invertebrates. Even those whose diet consists mainly of
seeds often feed their young in the nest on insects. Amongst
the mammals, pigmy shrews feed almost entirely on invertebrates, whilst wood mice include a fair proportion of insects in their diet. Fishes, common lizards and palmate newts
depend on invertebrates for their survival. The island has

not only retained fascinating Hebridean populations of animals and plants but also provides opportunities for management aimed at increasing the levels of biological activity. One of the most important and painstaking tasks as forests reappear on the barren hills of Rhum is the measurement of change in invertebrate communities.

Accounts of the invertebrate fauna of Rhum extend back to the last century. The earliest published records of insects recorded from Rhum were of water beetles collected by a visiting botanist, Symington Grieve, in 1884 and sent to B. H. Grimshaw in the Edinburgh Natural History Museum for identification. However, until 1960, the only person who had attempted to produce detailed observations on the island's insects was Professor J. W. Heslop-Harrison of Durham University who led regular expeditions to the island between 1938 and 1957. He concentrated his efforts on botanical surveys but also collected information on butterflies and moths, beetles, and gall-making insects which included flies, aphids and gall wasps. Many of the insects he recorded were subsequently confirmed from the localities in which he found them but a few rarities have never been subsequently recorded. He did not keep voucher specimens and doubts have been expressed over the authenticity of his records, including those of the large blue butterfly *Maculinea arion* (L.) (now considered to be extinct in Britain) and the slender scotch burnet moth *Zygaena loti* (D. & S.) which today is confined to the Isle of Mull and the Isle of Ulva, although it was originally recorded from mainland Argyll. A specimen of the slender scotch burnet reputed to have been one of the Professor's Raasay insects was seen by G. Tremewen of the British Museum (Natural History), who considered it to be a continental form and not the Scottish subspecies *Scotica*. The coleopterist, F. Balfour-Brown, mentioned twenty-eight species of water beetles collected from Rhum and sent to him for identification. Several of these insects, taken from 'a mountain pool on Rhum' turned out to be *Hydroporus foveolatus* which had not previously been recorded in the British Isles.

The Heslop-Harrison Rhum rarities will, perhaps, remain a mystery forever, but I shall continue to look out for the Large Blue butterfly in the Inner Hebrides.

Since 1960, the island's invertebrates have been the

subject of studies by several visiting scientists. A team of entomologists was invited to Rhum in the early 1960s with the aim of producing a basic catalogue of insects – including information on dates, localities, relative abundance and for selected species information on food requirements. Ten entomologists, each specialising in one or more orders of insects, visited the island for two weeks at different seasons in each of the four years 1960 to 1963. In spite of the fact that collecting was seriously hampered by rain for more than half the time, these initial surveys provided a good baseline and a list of 1,722 species was produced. In 1969, when the findings of the initial surveys were published, entomological surveys were resumed and six of the participants of the earlier surveys returned along with four specialists in groups which had not been studied in detail. Special attention was paid to the colonisation by insects of the woodland restoration enclosures. Subsequently, additional information was gathered as a result of visits to the island by other entomologists and by resident Nature Conservancy staff.

In 1982, the first catalogue, including all authenticated records up to 1980, was produced. The surveys produced many exciting records. For example, no less than ten species of aphids not hitherto recorded in Scotland, three of which were new to the British Isles, were recorded on the island in 1969. Because of the inhospitable nature of the climate and terrain, Rhum would not be expected to support a rich insect fauna and yet a catalogue of 2,158 species of insects has now been produced, representing ten per cent of the total British insect fauna. Table 5.1 shows the numbers of species recorded in the fifteen orders studied and, today, although the island's invertebrate catalogue is still far from complete, the insect fauna of Rhum is probably more fully documented than that of any other Scottish offshore island.

The larger Orders of insects, including the shield bugs, pond skaters, frog hoppers, aphids and scale insects (Hemiptera), the beetles (Coleoptera), the two-winged flies (Diptera), butterflies and moths (Lepidoptera) and the ants, wasps, bees, ichneumon flies and allied insects (Hymenoptera) have all received a great deal of attention and yet with each visit common and widely distributed species are added to the insect catalogue and it seems clear that, for most orders, the basic list is still incomplete. The most intensively

Table 5.1. The invertebrates of Rhum.

Order	No. of confirmed species	Percentage of British Fauna
Springtails (Collembola)	15	5
Mayflies (Ephemeroptera)	15	35
Dragonflies (Odonata)	10	22
Stoneflies (Plecoptera)	12	34
Grasshoppers (Orthoptera)	4	13
Earwigs (Dermaptera)	1	17
Booklice, etc. (Psocoptera)	10	20
Shield bugs, pond skaters, etc. (Hemiptera-Heteroptera)	84	16
Frog-hoppers, leaf-hoppers (Hemiptera-Homoptera: Auchenorhyncha)	67	19
Aphids, scale insects, etc. (Hemiptera-Homoptera: Sternorhyncha)	102	12
Alderflies, lace wings, etc. (Neuroptera)	10	17
Caddisflies (Trichoptera)	43	22
Butterflies and moths (Lepidoptera)	448	22
Beetles (Coleoptera)	523	17
Ants, bees, wasps, ichneumon flies, etc. (Hymenoptera)	235	5
Flies (Diptera)	550	9
Fleas (Siphonaptera)	8	14

surveyed group of insects is the Lepidoptera, which has been studied by resident entomologists over the years as well as visiting specialists, but populations of moths which have been overlooked over the years have been discovered as recently as 1980. The latest addition to the Lepidoptera list is the elephant hawk moth *Deilephila elpenor* which appeared at Kinloch in June 1982. In the past decade this species has extended its range northwards into the West Highlands and islands. In addition to the Rhum record it has, within the past five years, been found on Lismore, Colonsay and Canna. Of the larger Orders, the Hymenoptera requires much more attention before a truly comprehensive list is forthcoming. Many of the smaller parasitic wasps have been studied in depth by visiting specialists, but only a few of the more conspicuous ichneumon flies are recorded.

Inevitably, some orders containing small and inconspicuous insects have been overlooked. No records have been gathered of insects belonging to the orders Procura and Thysanura (bristle tails). In a study linking breakdown of litter in moorland soils with invertebrate fauna, twenty Colembola (spring tails) were recorded. No information has been gathered regarding thrips, nor biting or sucking lice parasitic on mammals or birds. Fleas, on the other hand, have received some attention, resulting in the discovery of a species new to science in the form of *Ceratophyllus fionnus* extracted from Manx shearwater nests on Hallival.

Crustaceans inhabiting Rhum's freshwater lochs have been surveyed, too. Short lists of earthworms, land and freshwater snails and slugs have been produced and fifty-eight species of spiders have been recorded, but far more information is required before a comprehensive checklist of these groups can be produced.

Aquatic Insects

Rhum is a wet island and blanket bogs cover large tracts of ground. Drainage is impeded by the uneven configuration of the land and marshy flushes and valley bogs are common. There are numerous stagnant peat pools, strongly acid in reaction, brown in colour and deficient in nutrients and oxygen. Larger pools and lochans occur from sea level to 350 m and varying in size from 0.2 to 2.42 ha in extent. The water in these is less acid and much clearer, but apart from the sea level loch at Papadil, the flora and fauna they support is very restricted. The richest populations of aquatic insects probably occur in Papadil Loch, fed by a stream which passes through a small plantation of deciduous trees, including areas of alder swamp, established around the turn of the century. Kilmory, with its meandering river and marshy meadows influenced to some degree by blown sands, is also rich in aquatic insects.

In spite of the hostile aquatic conditions, Rhum supports an interesting range of insects which breed in the water. The four species of damsel fly and ten of dragonflies that breed on the island are typical upland species. The rarest of these is the hawker dragonfly known as the blue aeschna *Aeschna caerulea,* which is confined to Scotland in the British Isles and is local in distribution. A small pool behind the raised beach at Harris, which is strongly influenced by salt spray, supports a surprisingly large number of dragonflies and this is the only locality to date for the water scorpion *Nepa cinerea.* The same pool supports a population of the china mark moth *Nymphula nymphulata* whose caterpillar forms a protective case made from leaves and feeds on the undersides of pond weeds.

The water beetles of Rhum are also predominantly found in the uplands. The largest of these is *Dytiscus semisulcatus,* whose voracious larva preys upon the tadpoles of the palmate newt, the only amphibian present on Rhum. Eleven species of water boatmen are recorded from the island and other water bugs including the Salidae, Velidae and Gerridae (water skaters) which are mostly northern and upland species.

Sixteen species of mayflies, twelve stoneflies and forty-three caddis flies, all of which start their lives as nymphs or larvae in the water, have been recorded from the island. Most of these are sparsely distributed, but some of the commoner species of mayfly swarm on still summer days and the commoner species of caddis fly appear in large numbers in lighted windows or in a mercury vapour trap set for moths. Many two-winged flies are aquatic or semi-aquatic in the larval stage. Very little research however has so far been devoted to some of the families whose larvae are truly aquatic e.g. Chironmidae. There are, however, craneflies, droneflies and mosquitos which start their lives in pools and, of course, the ubiquitous Rhum biting midges whose larvae live in the peat bogs.

Mountain and Moorland Insects

Three-quarters of the surface area of Rhum is covered by moorland and wet heathland vegetation or bare rock, screes and erosion terraces. Extreme exposure to the fierce winds which sweep from the Atlantic and periods of incessant rain limit the diversity of invertebrates which exist there. Yet some species appear in profusion. The magpie moth *Abraxas grossulariata* (L.) subsists on currant or gooseberry bushes in the south but in the west and north of Scotland it has adapted to moorland conditions where it feeds on heather. On Rhum, this moth occurs in such large numbers that many thousands of caterpillars fall into streams and pools from rank, over-hanging heather and drown. Brown trout have sometimes been caught in the hill lochs, their stomachs packed with these caterpillars. Other moths which occur in profusion over the moors include the heath rustic *Xestia agathina* and the true lovers knot *Lycophotia porphyrea*.

Several rare upland species of insects occur on the ul-trabasic mountains of Barkeval, Hallival and Askival. They include two ground beetles *Leistus montanus* (Steph.) and *Amara quenseli* (Schoen.) a small moth *Scrobipalpa murinella* whose larva mines in the leaves of mountain everlasting and a species of grass moth *Catoptria furcatellus* (Zett.) which occurs on grassy slopes at high altitudes. On Rhum, this species is confined to high level grasslands on the Manx shearwater colonies of Askival. Another rare leaf-miner *Nep-ticula dryadella* is found on mountain avens growing on the limestone outcrops of Monadh Dubh.

In late April, hairy caterpillars appear from the cover of heathland vegetation. Fat brown fox moth larvae which have developed into full-size feeding on heather the previous sum-mer and have spent the winter tucked away in the vegetation crawl up to bask in the warmth of the brief, bright intervals. The young caterpillars of the northern eggar are soon to be seen amongst the heather on which they, too, feed through the summer and pupate in the autumn. Tufted caterpillars of the drinker moth which have hibernated amongst the dead moorgrass appear in prodigious numbers to feed on the young shoots of *Molinia*. Where snowpatches remain on the hills, they sometimes fan out over the melting snow, drinking the droplets of water on the surface. At this time, too, the cuckoos return from Africa. The cuckoo appears to be the only bird in Britain which feeds on these hairy caterpillars and it is not surprising that it is a common bird over the moorlands of Rhum, since a ready supply of hairy caterpillars is available throughout the spring and summer.

Coastal Insects

Many insects are confined to the rocky coast of sand dunes or to the seashore itself. Some species of rare beetles are found only on the drift zone on the beach at Kilmory. The dunes and, behind them, the herb-rich links are sometimes swarming with a species of ladybird – *Coccinella undecimpunctata* var *boreolitoralis* (Don.) which seems to be confined to sand hills on northern coasts where no other species of ladybirds occur. Ladybirds feed on aphids and it is not surprising that these dunes and links also support the most interesting range of aphids to be found on the island, each one dependent upon a particular herb in the sward. Several species are confined to birdsfoot trefoil or kidney vetch, others to wild thyme. One species, new to Scotland, was found on buckhorn plantain. Most interesting of all the aphids is *Hyperomyzus thorsteinni* found on eyebright. This species had been described by Dr Stroyan from Iceland and it was he who recorded it from Kilmory as a new species to the British Isles.

The grayling butterfly is usually confined to rocky outcrops by the shore and to the sand dunes. Also a number of moths, including the white line dart, the archers dart, the coast dart and the short wainscot are characteristic species of coastal dunes or herb-rich grasslands. Perhaps the rarest of the coastland moths is the grey *Hedena caesia* which, in Scotland, is confined to a few islands in the Inner Hebrides where the caterpillar feeds on sea campion. On Rhum, it inhabits the coastal cliffs in the south and west of the island. The dew moth, an attractive, yellow, day-flying insect, also occurs in warm suntraps below coastal cliffs where its caterpillar feeds on lichens by the rocky shores.

In April, the wingless females of the belted beauty moth *Lycia zonaria* can be seen laying their eggs on lichen-covered rocks or pieces of driftwood near the beach at Kilmory. These are curious fat, hairy little insects, somewhat resembling grey, wingless bees. The males are handsome grey and white moths and they fly in the early evening from one female to another (figure 5.1). On hatching, the caterpillars crawl out over the grasslands, feeding on a wide range of flowering plants. In Scotland, this insect is largely confined to the Hebrides, occurring on only one mainland locality in Argyll.

One of the most interesting groups of moths which occur in the Hebrides are the burnets. They are spectacular red and dark green day-flying insects, some species of which are very local in distribution. They are represented on Rhum

Figure 5.1. The belted beauty moth: (a) male; (b) female.

by the six spot burnet, *Zygaena filipendulae*, which is common and widely distributed, and the transparent burnet, *Z. purpuralis*, which is found on sunny, south-facing slopes usually near the coast. The small slug-shaped caterpillars feed on wild thyme and the Rhum colony is largely confined to the south and west of the island. Apart from Rhum, the moth is found on several other islands in the Inner Hebrides, on Sanda Island, Kintyre and at a few mainland localities around Oban.

Woodland Insects

Woodland is the richest of all terrestrial invertebrate habitats and the policy woods at Kinloch support the greatest range of insects present on Rhum. These woods were created only eighty years ago in an almost treeless environment (see Chapter 4) and yet over twenty species of arboreal Heteroptera and a similar number of aphids are now established. All these rely on the leaves of particular trees and shrubs from which they suck the sap. Over one hundred and thirty moths are recorded whose caterpillars feed on the leaves of trees and bushes. One species – the lunar hornet moth, which closely resembles a wasp – spends its larval stage boring into the stems of willow trees. Woodland beetles such as the pine weevil and several species of bark and wood-boring beetles have gained a footing. One long-horned beetle which breeds in pine stumps, *Asemum striatum*, appeared in the 1960s and is now well established. Numerous beetles which live in bracket fungi or in toadstools on the forest floor are also well ensconced. Woodland sawflies which rely on particular trees and shrubs are resident, and the giant woodwasp *Uroceros gigas* (L.) which is not a true wasp but belongs to the sawfly family, is also firmly established. This is an insect of coniferous forests whose ovipositor consists of a highly efficient drill so that it can lay its eggs deep into solid timber.

With woodland insects have come their parasites and predators. The persuasive burglar *Rhyssa persuassoria* is the largest of all British ichneumon flies. It is a parasite of the giant woodwasp and can be seen searching over the tree trunks in which the wood-boring larvae occur. Once located, the ichneumon pierces the bark with its long, delicate ovipositor and deposits an egg in the horntail larva within its tunnel. The ichneumon larva then enters and feeds internally on the woodwasp.

Among the many arboreal moths which have become established in the Kinloch woods are a few whose larval food plants are exclusively non-native trees. Of these, a minute leaf-miner known as *Phyllonorycter maestingella* causes crimpling and browning of the leaves of beech. It is a woodland insect associated with a tree which is not a native species, established in a fairly remote island and yet no less than six species of hymenopterous parasites have been recorded preying on the larvae of this moth.

It is the native trees and shrubs which support the greatest number of species of woodland insects and although these are well represented in the Kinloch woods, there is only one mature oak tree. Since oak supports a richer fauna than any other tree in Britain, its more widespread use in recent plantings will undoubtedly result in the appearance of new woodland insects by immigration. Already several species of gall wasp have established themselves in the woodland restoration plots at Harris and Kilmory.

Fragments of scrub surviving in isolated situations support some woodland insects including the northern winter moths, of which the females are flightless. These scrub patches and the Kinloch woodlands now provide effective dispersal centres for woodland insects to colonise the new plantations of native Scottish trees and shrubs now being created over extensive areas in the north and east of the island. Changes are also in evidence in the bird populations as the woodland invertebrates on which many of them feed establish themselves.

Flightless Insects

Of the 523 beetles recorded from Rhum, over seventy belong to the family Carabidae (ground beetles) many of which are entirely flightless. They are ground living insects not at all well adapted for accidental transport across the sea. How did these insects get to the island in the first place? How long have they lived cut off from other populations of their own kind? Have they developed different characteristics through living in island isolation? All these questions have yet to be answered.

Like the northern winter moth, the females of several other species are flightless. The belted beauty *Lycia zonaria*, which inhabits coastal habitats, the mottled umber *Erannis defoliaria*, the scarce umber *Agriopis aurantiaria*, the dotted border *A. marginaria* and the common winter moth *Operophtera brumata* all have flightless females and their larvae usually live on deciduous trees and shrubs. On Rhum the dotted border and common winter moth have become adapted to moorland conditions, where the caterpillars live on heather, in addition to having thriving woodland populations in the Kinloch Castle policies. One flightless bug *Lamproplax picea* recorded at Kinloch is unexpected since it is otherwise found only in ancient forest relics in southern England.

Parasites and Bloodsuckers

It is said that, in the old crofting era, a form of punishment for wrong-doers in the Small Isles was to tie them out naked on a midgy night. Anyone who has experienced the viciousness of the Rhum midges on a warm, moist, summer evening will realise the agonising torment such a wrong-doer would be exposed to. The aggressiveness of biting flies on Rhum has to be felt to be believed. Between May and September the midges *Culicoides impunctatus* and *C. heliophilus* occur in unbelievable numbers.

In June and July, tabanid flies of five species are equally persistent in their pursuit of warm-blooded animals. The commonest of these are the clegs (*Haematopota* spp) of which two species occur: *H. crassicornis*, which is northern in distribution and, closely similar in appearance, *H. pluvialis*, which is generally more abundant in southern England. The relative abundance of these two species on Rhum is not fully understood but the latter seems to be active later in the summer. Other Tabanidae recorded are *Chrysops relictus*, a handsome insect with yellow and black markings, pictured wings and irridescent eyes, and two species of *Hybomitra* known locally as the green-eyed monsters: *H. bimaculata* and *H. montana*. All members of this ferocious family lay their eggs in compact masses on the stems of plants growing in boggy places. The larvae, which live in the peat or under stones in wet places, are carnivorous, feeding on small worms and the larvae of other insects.

The sheep tick *Ixodes ricinus* is another troublesome bloodsucker occurring over the moors. It will attach itself to deer, ponies, goats, human beings and birds, embedding its head into the skin and gradually becoming bloated with blood until it resembles a bean. Ticks will even feed on cold-blooded creatures such as common lizards and to sit on a flowery knoll at Papadil or Inbhir Ghil is a hazardous pastime. Within a few minutes hundreds of ticks, eager for fresh blood, can find their way to all parts of the body.

A number of flies present on Rhum are parasitic upon red deer. The deer ked, *Lipoptena cervi*, has wings when it first emerges from the pupa but, once it has reached its host, the wings are shed and the insect takes on a flattened spider-like appearance and spends the rest of its life on the deer, hiding amongst the thick hair. The deer warble fly *Hypoderma diana* is widespread and common. The females glue their eggs to the hairs of the deer. On hatching, the larvae crawl down the hair and penetrate the skin. During the next two months, they work their way through the chest and abdominal cavity, sometimes entering the tissues of the gullet. They then move to the back of their host, positioning themselves just beneath the skin. After making a breathing hole in the skin, they become surrounded by pus and the resulting swelling is known as a 'warble'. Here they remain for another two months. When fully developed, the larvae emerge through the breathing aperture and drop to the ground where they pupate, the adult insect emerging about a month later.

Another large fly is the nostril fly *Cephenemyia auribabis* which resembles a bumble bee. Demented deer are some-times to be seen pushing their noses down into the peat in a vain attempt to prevent these insects from gaining access to their nostrils. Nostril flies are viviparous, depositing living larvae into the nose of their host. The larvae find their way into the frontal sinuses and sometimes into the windpipe, remaining attached by large mouth hooks in the unfortunate deer for nine months. When fully fed, the fat corrugated maggots relinquish their hold and are sneezed out by the deer onto the ground where they pupate amongst the vegeta-tion. The adult life is short, the flies being on the wing in June and early July.

In addition to insect parasites the red deer on Rhum are sometimes infested with liver flukes. Several species of parasitic worms also subsist in their lungs, stomachs and intestines. In one stag, shot on Askival in the early 1960s Dr

A. M. Dunn discovered a nematode new to science and named it *Trichostrongylus askivali*.

The Rhum ponies also have their own parasites. The horse bot fly *Gastrophilus intestinalis* lays its eggs on the tips of hairs on the shoulders and flanks. As they groom one another, the ponies take these eggs into their mouths and the larvae live internally, attached to the stomach wall.

Little information is, as yet, available on the body lice of deer, wild goats, ponies and cattle nor on bird lice and it is hoped that future invertebrate studies will include not only these important groups but extended studies of the full range of mammal parasites with a special emphasis on the species which live on or in red deer. Studies on these important invertebrates should not be confined to Rhum. Many of them are confined in Britain to the Highlands and Inner Hebrides but apart from this little is known about distribution within their geographical range.

Future Invertebrate Studies

A number of species of butterflies and moths are represented in the Hebrides by distinct forms or sub species. There is considerable scope on Rhum for research into genetical variation in sedentary populations. Some of the flightless insects recorded on Rhum would seem to be first contenders in these studies. Changes to the habitat and management of Rhum provide important opportunities for studying the dynamics of invertebrate communities. The reintroduction of hill cattle to Rhum in the 1970s has probably affected the distribution of certain insects, particularly dung feeding beetles and flies in the south and west. Dramatic changes are taking place in the Rhum landscape as woodlands become established over extensive areas which were formerly treeless particularly in the north and east. Future invertebrate research should be devoted not only to filling in the gaps in our knowledge of endemic populations and migrant species but also to changes brought about by manipulation of grazing animals and by woodland restoration. Particular attention needs to be paid to insect colonisers, taking careful note of the arrival and establishment of new immigrant species. Great changes in invertebrate populations can be expected and an appraisal of these changes should be the main aim of future invertebrate surveys.

Further Reading

Fryer, G. & Forshaw, O. (1979) The freshwater crustacea of the Island of Rhum, a faunistic and ecological survey. *Biol. J. Linnean Soc. 11*, 333-67.

Steel, W. O. & Woodraffe, G. E. (1969) The entomology of the Isle of Rhum National Nature Reserve. *Trans. Soc. British Entomol. 18(6)*, 91-167.

Wormell, P. (1977) Woodland insect population changes on the Isle of Rhum in relation to forest history and woodland restoration. *Scottish Forestry 31(1)*, 13-36.

Wormell, P. (1982) The entomology of the Isle of Rhum National Nature Reserve. *Biol. J. Linnean Soc. 18*, 291-401.

6

THE BIRDS OF RHUM

J. A. LOVE AND P. WORMELL

Seabirds

The earliest published reference to the birds of Rhum is by
Dean Munro in 1549, and includes the comment that 'many
solane geese are in the isle'. Nowadays the term 'solane
geese' refers to the gannet *Sula bassana* and it is likely that
the Dean was alluding to manx shearwaters. In 1703 Martin
Martin wrote that there are 'plenty of land and seafowl: some
of the latter, especially the puffin, build in the hills as well as
in the rocks on the coast'. Again Martin must have been
referring to manx shearwaters *Procellaria puffinus*. The colony of this species is almost certainly of great antiquity and
must have been known to the Vikings when they named the
mountain 'Trollval' which today has the greatest density of
breeding shearwaters. In the Faroes at least three placenames
incorporating the prefix 'Troll-' denote important shearwater
colonies.

Climbing during the daylight hours amongst the ultrabasic peaks of south-east Rhum one can be unaware that
these slopes harbour one of the largest and most fascinating
of Britain's seabird colonies. One can see countless burrows
but rabbits cannot be responsible, for there are none on the
island. They are in fact the nest holes of the manx shearwaters. The underlying rock has been broken down through
time by frost, wind and rain to form a deep sandy grit which
drains well and which the birds can easily excavate using
their sharp hooked beak and powerful webbed feet.

Recently estimated to number over 100,000 pairs, the
colony is the largest in Scotland and is unique in Britain in
that it occurs above 450 m, on the upper slopes of the highest

78

The Birds of Rhum

Figure 6.1. Peridotite slopes on the north face of Hallival colonised by manx shearwaters.

hills (figure 6.1). The conical peaks of Hallival and Askival with their connecting ridge and the rugged summit of Troll-val support the majority – especially the north and east facing slopes. There are also a few burrows to be found on Fionchra, Ruinsival and Bloodstone Hill.

The shearwaters of Rhum return from their wintering grounds off the coast of Brazil in the third week of March. On summer evenings, vast rafts may gather offshore. After the last glimmer of sunlight has faded, the birds fly in to their mountain-top colonies. Shearwaters pair for life and return to the same nest hole year after year. They are long-lived and some ringed on Rhum as adults in 1958 were still returning to the same burrow on Hallival in 1972. A single egg, the size of a hen's, is laid in each nest. The incubation and fledging periods are fifty-three and seventy-three days respectively, so that the birds remain on their breeding grounds from March until October.

The chick grows big and fat for the first fifty days of the fledging period, and is periodically fed on regurgitated, semi-digested fish brought in by one or both parents. It sometimes becomes so obese that the exit hole is too small for it to leave. At this time, the parents abandon the young and only when it has used up some surplus fat does it emerge on to the dark hillside to exercise its wings. It finally fledges when, in a suitable wind, it clambers on top of a convenient boulder and launches itself down the mountain to the sea.

On land, manx shearwaters are awkward and shuffle around on their wings and belly, pushing with the feet. They are vulnerable to predators and are only active above ground under cover of darkness. Nonetheless, golden eagles *Aquila chrysaetos* and peregrines *Falco peregrinus* prey on shearwaters. The eaglet in one eyrie situated within the shearwater breeding area is often fed on these seabirds. Gulls have never been seen hunting around the colony although ravens *Corvus corax* and hooded crows *C. corone* will eat carcasses or any eggs which may be kicked out of the burrows. The discovery of shearwater carcasses with their heads and legs missing was for a time puzzling until it was found in the 1960s that red deer sometimes kill and eat bony extremities, especially the head.

Coastal Seabird Counts

Seabird studies on Rhum have not been confined to the manx shearwater: regular counts are carried out of fulmar *Fulmarus glacialis*, shag *Phalacrocorax aristotelis*, auks and gulls: the principal colonies are shown in Figure 6.2. In 1881 Harvie-Brown wrote that 'the whole circumference of Rhum may be looked upon as one vast colony of gulls, the herring gull outnumbering the lesser black-backs in the west coast but the lesser blackbacks being most abundant in the north-east'. In 1956 W. R. P. Bourne estimated 1,000 pairs along the western shores, but there are fewer today. Along the east coast, between Kilmory and Papadil, numbers of herring gulls *Larus argentatus* fluctuate between 237 and 536 pairs, and between thirty-six and sixty-four pairs of lesser black-backs *L. fuscus* breed there too. Scattered pairs of greater blackbacks *L. marinus*, occur all around the coast with about twenty pairs on one small sea stack called, appropriately, Stac na Faoileann. Common gulls, *L. canus*, breed in small numbers only, twenty-five pairs nesting regularly at Samhnan Insir and Welshman's Rock. Other small colonies were established in the 1960s at Kilmory, Camas Pliasgaig, Port na Caranean and Harris and, by 1965, the latter held about twenty-five pairs but fewer pairs have nested there in

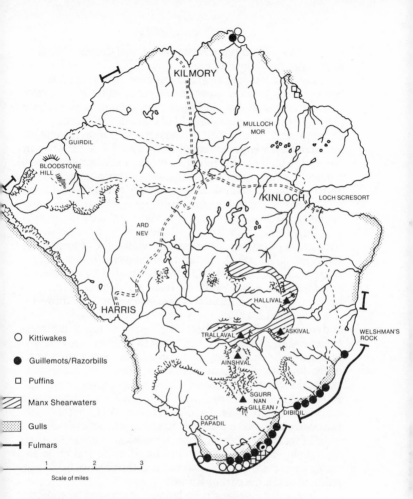

Figure 6.2. Distribution of Rhum's seabird colonies.

recent years. Kittiwakes, *Rissa tridactyla,* nest on inaccessible sea cliffs in dense colonies, which on Rhum are to be found only on the ledges of the Torridonian sandstone in the south-east, with a tiny colony at Samhnan Insir. It was in the north, however, that R. Gray first recorded them breeding in the 1880s. Numbers on Rhum have fluctuated in recent years between 800 and 1,600 occupied nests: in 1982 1,719 nests were counted but the population had dropped again to only 1,433 the following year.

Thomas Pennant's 'arctic gulls' or 'fasceddars' which he saw in July 1772 were in all probability arctic skuas *Stercorarius parasiticus*. Although they are now seen offshore in small numbers, none have bred on Rhum in recent years. Great skuas *S. skua* are also regular in the summer. Pennant also reported a large colony of terns round Loch Coire nan Grunnd, which he referred to as 'greater terns' although they are likely to have been either common or arctic *Sterna hirundo* and *S. paradisea*. Sandwich terns, *S. sandvicensis*, have been seen only once or twice. Any breeding in recent years however is in small numbers only and very irregular.

Razorbills, *Alca torda*, and guillemots, *Uria aalge*, are confined to sections of coastline where the Torridonian sandstone cliffs are sheer and offer convenient nest ledges, mainly around Dibidil and Papadil, with tiny colonies at Camas Pliasgaig and Samhnan Insir. Guillemots occupy the broader ledges or horiontal cracks where they stand row on row like penguins: razorbills prefer the more isolated cracks and crevices. The numbers of guillemots have increased from around 550 birds in the 1950s to 4,087 in 1983; razorbills fluctuate between 350 and 690 birds. Auks are notoriously difficult to count accurately, however, so these annual estimates serve only as an index of population change. Puffins, *Fratercula arctica*, which apparently once numbered several hundred pairs, have declined so that by 1976 only thirty-five pairs were counted: in recent years there has been a slight recovery to about sixty pairs. Black guillemots, *Cepphus grylle*, breed in narrow and inaccessible cracks on the rocky coasts and between twenty and thirty pairs have been counted along the east of Rhum.

The most dramatic population increase has been shown by the fulmar which first bred on Rhum around 1934 when twenty-two pairs had become established at Dibidil. In 1955 there were over 250 pairs, with a new colony establishing on Welshman's Rock, and by 1969 there were over 400 occupied nests along the south and east coasts of Rhum. The birds are prospecting new colonies each year and, in 1982, a total of 694 occupied nests were counted. In 1871 Gray claimed that there was 'a more extensive breeding place for leach's petrels *Oceanodroma leucorrhoa* on the island of Rhum situated on rough stony ground at the north-west end of a place called Braedinach'. Harvie-Brown failed to locate this colony however, as have we, searching the coastal screes at night using tape recorders and mistnets. In 1976 and 1977 a small number of storm petrels *Hydrobates pelagicus* were caught in nets at several localities around the coast and it is possible that a few may breed in some boulder screes.

Shags nest in scattered colonies among boulderfields all around the island: although a few hundred can be counted the total population may be much larger than this in some years. A few cormorants *Phalacrocorax carbo* and gannets may be seen offshore, the former mainly in the winter and the latter in the summer following the mackerel shoals.

Landbirds

Rhum is the largest of the four Small Isles and offers the greatest variety of habitats for birds. One hundred and ninety four species have been recorded on the island and among the eighty-seven species which have at one time or other bred on Rhum, more than fifty do so regularly: half of them are passerines occurring in the woodland around Kinloch. Chaffinches *Fringilla coelebs* and robins *Erithacus rubecula* are the commonest species with willow warblers *Phylloscopus trochilus* and blackbirds *Turdus merula* next, closely followed by goldcrest *Regulus regulus*, wren *Troglodytes troglodytes*, song thrush *Turdus philomelos* and dunnock *Prunella modularis* (figure 6.3). The abundance at Kinloch of some passerines such as treecreeper *Certhia familiaris* and coal tit *Parus ater* fluctuates according to the severity of the winter weather; other species such as the long-tailed tit *Aegithalos caudatus* may disappear altogether.

The history of the tiny blue tit *Parus caeruleus* population of the Kinloch Woods is better documented than most. In recent decades, as many as ten pairs have bred, although from 1961–63,in 1968 and 1969 they apparently died out altogether. These are years in which the January or February mean minimum temperatures fell near or below freezing. After the equally severe winter of 1977–78, however, the breeding population declined from eight to five pairs while after the 1981–82 winter the population remained stable at around seven pairs. Some influx from the mainland must have prevented the full effect of the hard weather and several

1) Chaffinch	43 pairs	12) Collared dove	6 pairs
2) Robin	43	13) Long-tailed tit	6
3) Willow warbler	28	14) House sparrow	6
4) Blackbird	28	15) Treecreeper	4
5) Goldcrest	21	16) Cuckoo	3
6) Wren	18	17) Woodcock	2
7) Song thrush	17	Hooded crow	2
8) Dunnock	14	Mistle thrush	2
9) Wood pigeon	10	Grey wagtail	2
10) Blue tit	9	18) Common sandpiper	1 pair
11) Coal tit	7	Spotted flycatcher	1
		Pied wagtail	1

Figure 6.3. Kinloch Castle policies' bird community, 1974.

of the immigrants stayed on to breed in subsequent summers.

In recent years as the plantings around Kinloch have matured, several pairs of chiffchaff *Phylloscopus collybita* have established and occasional pairs of greenfinch *Carduelis chloris*, bullfinch *Pyrrhula pyrrhula* and siskin *Carduelis spinus*. One or two pairs of spotted flycatcher *Muscipapa striata* and whitethroat *Sylvia communis* are present each summer and sometimes garden warbler *S. borin*, wood warbler *Phylloscopus sibilatrix* and grasshopper warbler *Locustella naevia*. Blackcaps *Sylvia atricapilla* frequently overwinter and may breed in some years. Whinchat *Saxicola rubetra* and reed buntings *Emberizza schoeniclus* are more frequent now, while on the open moor stonechat *Saxicola torquata* and wheatear *Oenanthe oenanthe* provide a welcome flash of summer colour replaced by the occasional snow bunting *Plectrophenax nivalis* in winter. The arrival of the cuckoo *Cuculus canorus* on Rhum has long been taken by the estate staff as a sign to plant the potatoes although in fact the bird's appearance has been remarkably consistent over the years while records have been kept: over forty per cent of annual first arrivals fall within only three days (around a mean of 27 April) although the actual arrival dates might range from 16 April (a surprisingly early date which occurred in 1984) to 5 May.

The Birds of Rhum

Despite the profusion of insects on Rhum, swallows *Hirundo rustica* are reluctant to linger at Kinloch and have bred only once or twice: single pairs usually breed at Harris and/or Kilmory. Flocks of martins *Riparia riparia* and *Delichon urbica*, and less frequently swifts *Apus apus* often turn up on passage. Almost without fail, one or more turtle doves *Streptopelia turtur* appear at Kinloch each May or June with as many as six being seen around the forest nursery in May 1979. Collared doves *S. decaocto* were first recorded on Rhum in 1961 from which time they have increased to some ten pairs: their numbers are sometimes reduced in cold winters. Wood pigeons *Columba palumbus* are common at Kinloch and a few now frequent the Kilmory tree plots. Rock doves *C. livia* are to be seen on the rocky coasts.

Hooded crows *Corvus corone* abound all over the island, nesting on crags or in trees. As many as sixty may flock at Kinloch for the winter, where they often feed on shellfish, cracked open by being dropped on any convenient flat rock near the shore or even on the rocky hillside behind Kinloch Woods. Perhaps a dozen pairs of ravens *C. corax* breed, but choughs *Pyrrhocorax pyrrhocorax* have long been extinct. In 1871, Robert Gray concluded that choughs 'no longer bred on Rhum' and the last recorded nesting on Eigg was in 1886. Rooks *C. frugilegus*, carrion crows *C. corone corone* and jackdaws *C. monedula* are only occasional winter visitors to the island.

Wintering blackbirds and starlings ringed on Rhum have been recovered in Scandinavia, but perhaps the most remarkable recovery was a blackbird caught on the island in November 1961. It was recaptured in northern Germany in March 1964 and returned to Rhum the following winter before finally being found dead in Denmark in April 1965.

Gamebirds

Neither partridge *Perdix perdix* nor pheasant *Phasianus colchicus* managed to establish on the island despite periodic releases of captive-reared birds in the last century and during the first two decades of this century. The ptarmigan *Lagopus mutus* bred on Rhum until the 1820s but is now extinct. In 1889, some two hundred red grouse *Lagopus lagopus* were released and the estate game books record up to eight hundred brace being shot in following years. Thereafter the bags

85

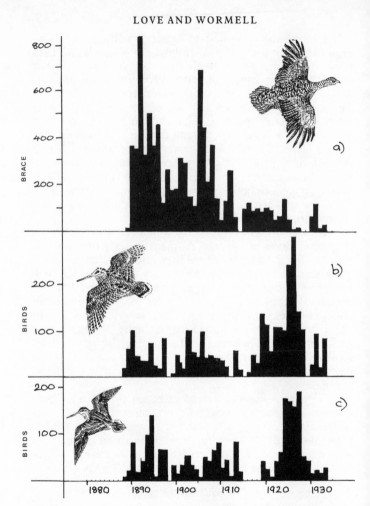

Figure 6.4. Annual bags of (a) red grouse, b) woodcock and (c) snipe on the Isle of Rhum, 1888-1934 (from the Bullough game books, Kinloch Castle).

declined with only a few being shot in the 1930s (figure 6.4).
Because of the poor heather cover, Rhum now supports
fewer than one hundred pairs of grouse and their breeding
success fluctuates annually according to weather conditions.
Quail *Coturnix coturnix* have been recorded on the island on
three occasions.

It is sad to encounter corncrakes or landrails *Crex crex* in
the game books of the Bullough estate: no fewer than eleven
were shot in September 1908. In 1934 six corncrakes were
heard calling amongst the derelict gardens and lawns of
Kinloch Castle. In the late 1960s, the numbers declined until
by 1975 none could be found. In subsequent years, one or
two have turned up but they rarely persist in calling for more
than a week or two and probably do not now breed.

Wading Birds, Wildfowl and Divers

Considerable numbers of snipe *Gallinago gallinago* and
woodcock *Scolopax rusticola* were also shot until the 1930s,
mostly passage birds during the winter months (figure 6.4).
Today common snipe are frequently flushed all over the wet
moorland of the island though it is difficult to assess whether
their winter numbers now match the bags achieved earlier
this century. Migrant woodcock still occur in late autumn
and each summer at dusk roding birds can be seen in the
woods around Kinloch.

Oystercatcher *Haematopus ostralegus*, ringed plover
Charadrius hiaticula and common sandpiper *Tringa
hypoleucos* breed commonly around the coast while the hills
support large numbers of breeding golden plover *Pluvialis
apricaria* of which a small flock usually winters near the shore
at Harris. No dunlin *Calidris alpina* or greenshank *Tringa
nebularia* breed on the island nor do redshank *Tringa totanus*
or lapwing *Vanellus vanellus*. The few pairs of nesting curlew
Numenius arquata, though slowly increasing, are still local in
their distribution and small flocks winter at Harris. Several
other waders, some of them comparative rarities, occur on
passage but with the area of sand and mudflats being small
and localised their numbers are few.

With the exception of eider *Somateria mollissima*, which
are common all round the coast of Rhum (especially at the
mussel beds in Loch Scresort), seaducks are vagrants to the
island – even common scoter *Melanitta nigra*, goldeneye

Bucephala clangula and long-tailed duck *Clangula hyemalis*.
Surf scoter *M. perspicillata*, velvet scoter *M. fusca* and king
eider *S. spectabilis* have been recorded on one or two occasions. Several pairs of red-breasted merganser *Mergus serrator*
breed annually around Loch Scresort while broods of mallard *Anas platyrhynchos* and teal *A. crecca* are usually seen on
some of the hill lochs. A pair of shelduck *Tadorna tadorna* are
regular nesters at Kilmory and in some seasons another pair
may breed on the shore at Shellesder.

Whooper swans *Cygnus cygnus*, grey geese *Anser* spp
and barnacles *Branta leucopsis* are frequently seen on passage,
whilst stray parties of both Canada geese *Branta canadensis*
and pale-bellied Brent *Bernicla hrota* have been recorded. In
1980 a pair of greylags *A. anser* attempted to breed but
disturbance caused them to desert. In 1982 two broods
hatched and six geese – probably a family party – stayed over
winter: three broods were reported in 1983. Graylags have
also recently become established on Canna and Muck.

Great northern divers *Gavia immer* are regularly seen
offshore until late spring but only red-throated divers *G.
stellata* breed on Rhum. From mid-March pairs or individuals can be seen flighting from feeding grounds at sea to their
chosen breeding lochs inland. Persecution by man used to
seriously reduce their numbers and, as recently as 1934, no
red-throats bred on the island. In 1950 a nest was found and,
by 1974, as many as fourteen pairs attempted to breed. Their
success varies, perhaps the most significant factor being the
flooding of the nests in wet summers, although crows are
quick to snatch unattended eggs if the adult is put off by
fishermen or walkers. The little grebe *Tachybaptus ruficollis*
is the commonest vagrant grebe but in common with
moorhen *Gallinula chloropus* (nine records), coot *Fulica atra*
(four records) and water rail *Rallus aquaticus* (eight records)
they do not find the island's mainly acid lochs suitable for
breeding. Herons *Ardea cinerea* are commonly seen along the
shore and sometimes venture inland to lochs and burns: one
or two pairs have nested in the trees around Kinloch.

Birds of Prey

Since there are relatively few small mammal species on Rhum
there are correspondingly few birds of prey. Six species are
mere vagrants and they include a rough-legged buzzard

Buteo lagopus and goshawk *Accipiter gentilis*. Ospreys *Pandion haliaetus* are more frequent (from May to July usually) and hen harriers *Circus cyaneus* are regular on spring and autumn passage. Barn owls *Tyto alba* have been seen on Rhum only twice although in 1891 a pair bred in a sea cave on Eigg. Tawny owls *Strix aluco* still occasionally breed on Eigg but are extremely infrequent on Rhum – surprisingly so since the woodland at Kinloch would seem suitable for them. Long-eared owls *Asio otus* do nest intermittently as do short-eared owls *A. flammeus* but the absence of voles means that these owls are rarely seen abroad by day.

Two or three pairs of merlins *Falco columbarius* breed on Rhum although recently they have become somewhat less numerous. The island supports a similar number of kestrels *F. tinnunculus* and also two or three pairs of peregrines *F. peregrinus*. In 1970, however, peregrines had ceased to breed on the island doubtless suffering from the effects of toxic chemicals derived from their seabird prey – a factor which had contributed to a nationwide decline of peregrines at that time. Happily in 1980 a pair re-established and reared one chick and now they have almost recovered their former abundance.

Last century, both golden eagles *Aquila chrysaetos* and white-tailed sea eagles *Haliaeetus albicilla* bred on Rhum but, like all predators at that time, they were victims of a relentless campaign of persecution by shepherds, gamekeepers and egg-collectors. In the late 1820s, one of the Rhum shepherds shot five sea eagles in one day whilst in a single year (1866) one local gamekeeper shot no fewer than eight sea eagles. When John Bullough bought Rhum in 1886, his staff were encouraged to adopt a tolerant attitude to predators, but his son reverted to the more traditional approach: in 1907 his keeper shot an adult sea eagle and collected two eggs from its cliff eyrie. The widowed bird found a new mate but both eagles were shot two years later. By 1916, sea eagles were totally exterminated throughout Britain, the last nest being on the Isle of Skye, only ten miles from Rhum.

The golden eagle was also persecuted. On Rhum four were shot in 1924 and ten years later a party of visiting ornithologists could only find one individual on the entire island. At two roosts we have found ancient gin traps – one rusted solid in the set position! But in many parts of the

Highlands, golden eagles frequented mountainous districts remote from the activities of man and when persecution was relaxed during the world wars, they managed to stage a recovery.

When Rhum became a nature reserve in 1957, three or four pairs had re-established territories, two of them on or near the coast and in eyries formerly used by sea eagles. Each pair has several alternative nests although only one or two tend to be favoured. All eyries on Rhum are on cliff ledges, sometimes with a small tree growing in front. Golden eagles may indulge in aerial courtship on any fine day during the winter, the pair often soaring together for long periods. Display becomes more frequent in February and March, culminating in copulation which may take place near the nest, or one of the alternatives which may not in fact be used that year. Laying commences in mid-March although fifty-three per cent of clutches are laid during the first week of April. Incubation, by both sexes, begins with the first egg and lasts about forty-four days. The normal clutch is two eggs but, since they hatch two or three days apart, the younger eaglet usually dies, deprived of food or even killed by its larger sibling. The surviving eaglet remains in the nest for about ten or eleven weeks, usually making its first flight in late July. It commonly remains with the adults for several more months before becoming fully independent.

The breeding success of Rhum's golden eagles is comparatively low: on average only one chick is raised on the island per year. Unhatched eggs have been found to contain high residues of toxic chemicals, notably DDT and PCBs, probably derived from the gulls and fulmars being eaten by the eagles. It may be significant that of the island's four pairs of golden eagles, the one with the best reproductive output kills fewest seabirds and feeds principally on red grouse and carrion from goat and deer carcasses. It is, however, encouraging to note that since 1977 the reproductive rate success has nearly doubled with three eaglets being reared on the island in that year.

Since the last nesting attempts of sea eagles took place on or near Rhum it was appropriate that attempts to reintroduce the species should be made there. Few sea eagles have been reported in Britain since the extermination of its breeding population in 1916. Unlike the migratory osprey, which is a regular migrant to our shores and has recently re-established itself unaided, the more sedentary sea eagle appeared unable to do so. Two early attempts at reintroduction (in Glen Etive in 1959 and on Fair Isle in 1968) were unsuccessful: only seven birds were released and neither venture was sustained for longer than one year. The Rhum experiment was begun in 1975 by the Nature Conservancy Council and involved four eaglets taken from eyries in Norway. After a brief period of captivity on Rhum – during which time the only male unfortunately died – the young eagles were released. They readily utilised the food provided until they were able to fend for themselves. One was subsequently found dead under overhead power cables but the other two adapted well to the wild.

Had the programme then ceased, the reintroduction would have been doomed to failure but, with the assistance of the World Wildlife Fund, the Royal Society for the Protection of Birds and the Scottish Wildlife Trust, the Nature Conservancy Council has been able to sustain importing birds each year. Ten eaglets were received in 1976, four in 1977, eight in 1978, six in 1979, eight in 1980, five in 1981, ten in 1982 and ten more in 1983, in 1984 and 1985 – all from Norway. Three died in captivity but a total of eighty-two eagles have been set free on Rhum. Prior to its release, each bird was weighed and measured to determine its sex. Females are larger than the males and, weighing on average 5.5 kg, are about 1 kg heavier. The largest eagle tipped the scales at 7.25 kg and had a wingspan of nearly three metres. The sexes look alike although the males have a slightly darker plumage (figure 6.5). They also tend to be more highly strung and to have a higher pitched call. Over the years, we have received a slight excess of females – forty-three as opposed to only thirty-nine males.

Releases commence in August and tethered birds awaiting their freedom help to decoy the others back to feed nearby. As the project has progressed, the inexperienced juveniles have often benefited from the presence of older birds released in previous years. They learn by their example and at times are even provided with food by them. Never having been allowed to become tame or imprinted while in captivity, the young eagles readily adapt to a free-flying existence and experience little difficulty in finding food. One

Figure 6.5. First-year sea eagle calling, showing the dark plumage of the juvenile, dark grey beak and brown eyes contrasting with the adult.

male succeeded in catching a hooded crow which had strayed within his reach while still held on tethers. Initially, the released eagles depend upon the supplies of food specially provided and on carrion, but, within a few months, some are able to catch and kill fish and seabirds for themselves. Gulls and auks are favoured, although the eagles often chase crows. Away from Rhum, rabbits and hares are frequent prey.

Seven of the released birds – three males and four females – are known to have died shortly after release and other deaths may have gone unrecorded. It is particularly sad to note that two individuals were poisoned, one being a bird in its fourth year and almost old enough to breed.

We have amassed an encouraging number of sightings of nearly fifty different individuals. The birds are finding the Western Isles a very suitable habitat and their survival is high. Sea eagles attain breeding age around their fifth year of life, by which time they have acquired the bright yellow beak and the white tail of adulthood. It is often a source of confusion that immature sea eagles lack the white tail, while the juvenile golden eagle's tail has white on it. The tail of the latter is long and fan-shaped and terminates in a broad black band: the white together with white patches on the wings, is lost as the golden eagle reaches maturity. One useful, though not always infallible guide is that the golden eagle often soars with its wings tilted upwards from the shoulder in a shallow 'V' while the sea eagle usually holds its wings horizontal. Also, the sea eagle is larger with a wing span longer by half a metre or more, and is more vocal.

At least five of the released sea eagles have paired up and have established breeding territories in the west of Scotland. The aerial courtship of these birds involves a spectacular talon-grappling display and mutual cartwheeling accompanied with much excited yelping. During 1980 we observed several birds carrying sticks and the following year one pair built a crude nest close to an eyrie used by their own species a hundred years before. In 1982, another pair constructed a crude nest but again nothing came of this attempt. Activity became more intense in 1983 at this latter site and by early April, the female was incubating eggs. Unfortunately her mate was being courted in the meantime by a second female who also managed to lay a clutch of her own in the same nest! One clutch (probably two eggs) was broken in the process. Shortly afterwards, the new female ousted the first completely and, although incubation of the second clutch was sustained for about forty days, the two eggs failed to hatch. It is likely that they too had been damaged during the complex interactions of the trio of adults.

In 1983, we found the eyrie of another pair and the female, an immature, was incubating a single egg. Unfortunately, this was thin-shelled and broke – apparently not an unusual occurrence in such young birds. In the meantime yet another eyrie was located which seemed to have built up the previous year but we may never know whether it had been used or not.

In 1984, the trio of adults did not attempt to breed but two other pairs laid eggs. One clutch was abandoned midway through incubation – perhaps during a period of cold weather which forced the pair to go off in search of food – and the second clutch (only one egg) was carefully incubated for nearly fifty days but failed to hatch. The egg has been collected for analysis.

Four pairs attempted to breed in 1985 and one chick was successfully fledged. Two chicks were reared in 1986. No more importations are planned at present.

Other pairs have now established at suitable breeding localities so the prospects for young being reared in the near future remains high. It is hoped that these events will lead to the establishment of a viable population in the Hebrides where not long ago sea eagles flourished.

Further Reading

Bourne, W. R. P. (1957) The birds of the Isle of Rhum, *Scottish Naturalist 69*, 21-31.

Corkhill, P. (1980) Golden eagles on Rhum, *Scottish Birds 11*, 33-43.

Evans, P. R. and Flower, W. V. (1967) The birds of the Small Isles, *Scottish Birds 4*, 404-45.

Love, J. A. (1981) An island population of blue tits, *Bird Study 28*, 63-4.

Love, J. A. (1983) *The Return of the Sea Eagle*. Cambridge University Press.

Love, J. A. (1984) *The Birds of Rhum*. NCC publication.

Wormell, P. (1976) The manx shearwaters of Rhum, *Scottish Birds 9*, 103-11.

7

RED DEER

T. H. CLUTTON-BROCK AND F. E. GUINNESS

History and Natural History

The history of Rhum's red deer population is not unusual. In the sixteenth century deer were common on the island: in 1549, Dean Munro noted 'an abundance of little deire' and they were still numerous by 1750. In the second half of the eighteenth century, as the human population rose towards three hundred, the deer population became extinct. The *Old Statistical Account* (1796) links their disappearance to the destruction of the woodland. 'There were formerly great numbers of deer: there was also a copse of wood that afforded cover . . . While the wood throve, the deer also throve; now that the wood is totally destroyed, the deer are extirpated'. However, the ability of subsequent populations to live without woodland suggests that other factors were responsible for their disappearance.

Deer were reintroduced to Rhum from the mainland after 1845, when the bulk of the island's human population had left. During the first half of the nineteenth century when Rhum was developed as a sporting estate by the Bullough family, it carried a population of between 1,200 and 1,700 deer. A small number of stags from English deer parks were introduced in the hope of improving the stock and around forty stags and forty hinds were culled annually.

In 1934 and 1935, Frank Fraser Darling carried out the first detailed field study of the behaviour and ecology of red deer – in the forests of Dundonnell and Letterewe and Gruinard in Wester Ross. Darling spent the best part of two years watching deer and working with stalkers, eventually producing a monograph many years ahead of its time which provided a model for future research on deer throughout the world. In it, Darling synthetized the available information on the reproductive cycle, feeding ecology and social behaviour of hinds and stags. Few of his observations can be faulted and his book still provides the most readable account of the natural history of Scottish red deer.

But there are many important questions about red deer which Darling's book did not answer. Darling carried out his field study at a time when field ecology had not yet become a quantitative science, before accurate ageing techniques had been developed and before it was possible to catch and mark individual animals. At what age do hinds start to breed and at what age does their fecundity begin to decline? What are the natural lifespans of hinds and stags? How far do deer disperse from the area where they are born? How does the body weight of the two sexes change throughout the year? To what extent do they depend on different food plants in different seasons? How does population density affect feeding ecology and reproduction? Darling's book provided no precise answers to these questions, yet all are of fundamental importance in managing deer stocks.

The purchase of Rhum by the Nature Conservancy in 1957 provided an open air laboratory where these and other questions about the ecology of upland plants and animals could be investigated. In 1957, the island's 10,600 hectares supported a self-contained deer population of 1,600 and offered a site where the requirements of sport would be subordinate to those of scientific research. After 1957, the deer cull was increased to around one sixth of the adult population in spring; the grazier's herd of 2,000 sheep was removed; and the practice of regularly burning parts of the moorland vegetation to provide fresh growth ceased.

Since 1958, three principal research projects have used the island's deer population. Between 1958 and 1970, the Nature Conservancy's scientific staff (including V. P. W. Lowe, B. Mitchell and B. E. Staines) examined the demography and feeding ecology of the population, paying particular attention to topics closely related to deer management. Their work provided the first quantitative description of the ecology of red deer and now provides a scientific basis for deer management throughout Scotland. Between 1967 and 1972, scientists from the Department of Veterinary

Figure 7.1. Related hinds often look alike: (a) a 16-year-old hind, with her 2-year-old daughter (*right*). The two hinds closely resemble each other in ear shape and patterning as well as in the general conformation of their faces. (b) An elderly hind, her two daughters and two calves. The mother (*third from left*) has a pronounced black mark at the base of the tail, as does her calf (*far left*). In contrast, her grandson (*centre*) has a particularly light-coloured rump patch.

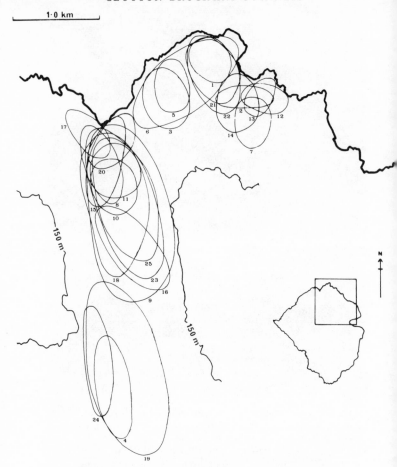

Figure 7.2. Core areas of matrilineal groups of hinds. Each
ellipse shows the smallest area which accounts for 65% of
the sightings of all members of a particular matriline. Numbers associated with each ellipse refer to the identity of the
matriline and do not reflect its size.

98

Red Deer

Medicine at Cambridge (including Roger Short, Gerald Lincoln, Fiona Guinness and John Fletcher) worked on the reproductive physiology of stags and hinds and the hormonal controls of breeding behaviour. Finally, since 1972, the Kilmory area has been used for studies of population regulation and of the causes of variation in breeding success in deer by another group from the Department of Zoology in Cambridge, including Steven Albon, Tim Clutton-Brock and Fiona Guinness.

Social Behaviour

On Rhum, as in most areas of the mainland, hinds are usually found in parties of two to twenty animals. These groups typically consist of females belonging to one or more kin groups: while sons disperse from their mother's home range between the age of two and four years, daughters adopt home ranges overlapping those of their mothers and live there throughout their lives, associating with their sisters, aunts and other matrilineal relatives more frequently than with unrelated animals (figures 7.1 and 7.2). Larger parties of hinds are usually made up of several small groups belonging to different matrilines and vary in membership from hour to hour. When they are in the same party, relatives are tolerant of each other's proximity, threatening and displacing each other less than unrelated animals. The home ranges of different matrilines overlap widely and there is no evidence of hostility or territoriality between neighbouring groups.

Compared to the matrilineal groups of hinds, stag groups are loosely structured. Young stags leave their mother's home range and begin to wander widely between the ages of two and four years. Gradually, they spend more and more time in one or more of the areas regularly used by other stags. Unlike hinds, stags with overlapping home ranges do not associate regularly with each other and stags are less tolerant of each other than hinds. Dominance rank is consistent but is not closely related to their breeding success during the rut or – as is sometimes claimed – to their antler size.

Stags remain in these loose, bachelor herds through the summer though mixed sex groups become more common as the grazing improves in late September. One by one, they leave their usual haunts and travel to their rutting grounds –

Figure 7.3. Two stags fight for a harem during the October rut.

which may be several miles away from their winter range. They appear to adopt the area where they first successfully hold a harem as their traditional rutting ground, returning to it year after year until they are finally displaced by a rival.

Fighting for harems is not uncommon and injuries regularly occur (figure 7.3). Our long-term records suggest that mature stags fight, on average, four to five times each rut and that, in each year, they have a one in twenty chance of being seriously injured.

Though fighting is necessary to win and hold a harem, it pays stags to fight as little as possible – because every time a stag fights he is risking permanent injury. Stags assess their rivals in the roaring contests that occur in situations where a fight is likely. The rate of roaring that a stag can maintain is related to his strength and fighting ability and by engaging in roaring matches before fighting, challengers can assess the capacity of their opponents, pressing home a challenge only where they have out-roared an opponent.

Reproduction

In red deer, as in many other seasonally breeding mammals, the timing of the mating season is very regular. On Rhum, stags first start to hold harems in the last week of September and mating activity reaches its peak in the second and third weeks of October. Only hinds in oestrus will allow the stag to mount. Oestrus usually lasts for less than twenty-four hours and will be repeated every eighteen days until the next spring if the hind is not fertilized. A single calf is born approximately eight months after fertilization (our estimates of gestation length are 234 days for female calves and 236 for males). Twins occur very occasionally – two cases have been known on Rhum in the last twenty years.

The strong seasonality of reproduction is associated with changes in the stag's reproductive organs. Unlike many mammals, red deer stags are only fertile during the autumn and part of the winter and their testes change in size throughout the year, increasing from a combined weight of around 50 g in spring to a combined weight of 150 g by late September. Following changes in testis size, hormone levels begin to rise in early September, peak during October and decline rapidly in November.

The antler cycle is controlled by seasonal changes in testosterone. After the winter solstice, testosterone secretion and spermatogenesis are suppressed by increasing day length, leading to antler casting in March. The loss of the old antler immediately triggers new growth and the dark velvety knobs of next year's antlers sprout from the old craters. Antler growth is rapid and is almost complete by the end of July when rising testosterone levels lead to the constriction of the blood vessels serving the antlers, to the withering of the skin covering them and eventually to the cleaning of the antlers. Casting, followed by new antler growth can be induced at any time during the winter by castration – and cleaning by the replacement of testosterone by implant.

Hormonal changes influence the stags' behaviour. After the rut, when stags rejoin their bachelor groups, a linear dominance hierarchy forms. An individual's rank within the hierarchy determines his access to food supplies and his social relations with his neighbours. The older and larger stags are the first to cast their antlers in the spring and this has an immediate effect on their success in social encounters. Without antlers, they are no longer able to displace younger animals that are still in hard horn and the hierarchy quickly becomes disorganised. As more and more of the stags cast and begin to grow new antlers, a clear rank order gradually re-emerges, based not on antler threats and clashes but on interactions in which stags stand on their hind feet and slap

Figure 7.4. In summer, two stags in velvet box at each other with their fore-feet.

at each other with their feet (figure 7.4).

Traditional knowledge has it that red deer are very long-lived: 'Thrice the age of a dog is that of a horse, thrice the age of a horse is that of a man, thrice the age of a man is that of a deer . . .' However, this is far from the case. In those parts of the island where the population is culled each year, over ninety per cent of both sexes have died by the time they are ten and, even in the Kilmory area where the deer are allowed to live out their natural lifespans, over ninety per cent of stags die before the age of eleven and over ninety per cent of hinds by the age of fourteen (figure 7.5).

Red Deer

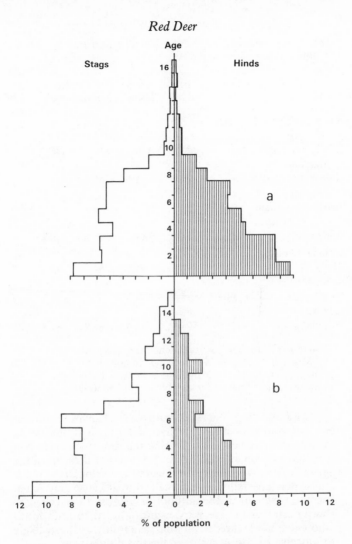

Figure 7.5. (a) Year class percentages in the red deer population of Rhum on 1 June 1957 reconstructed from subsequent deaths and estimated survivors up to May 1966 (after Lowe 1969). (b) Percentages of animals in different year classes in the North Block in 1971.

Table 7.1. Measures of reproduction and performance in different Scottish deer forests.

Measure	Glen Dye[4]	Glen Fiddich[1]	Glen-feshie[1]	Rhum	Inver-mark[1]	Scarba[2]
Density (per km²)	1.6	8.0	13.1	13.9	15.3	34.4
Adult sex ratio (hinds/100 stags)	198	137	140	116	556	157
Calf/hind ratio (calves/100 hinds)	45.8	30.7	34.6	38.1	41.5	39.2
Percentage of yearlings pregnant	64	20	0	0	43	0
Yearling hind larder weight (kg)	42.6	45.3	38.0	37.0	40.4	37.2
Percentage of milk hinds pregnant	94	92	44	51	83	39
Milk hind larder weight (kg)[b]	54.7	55.8	46.5	51.0	51.7	58.9
Stag larder weight (kg)[c]	78.6	105.8[a]	83.8	88.7	92.1	93.7

Sources: [1] Mitchell 1967; [2] Mitchell & Crisp 1981;
[3] Mitchell, Staines & Welch 1977; [4] Staines 1978.
[a] Stags fed in winter. [b] Milk hind larder weight 60% of live weight. [c] Stag larder weight 73% of live weight.

The breeding success of both sexes changes with their age. On Rhum, most hinds bear a calf for the first time either as two- or three-year-olds. During the rest of their lifespan they conceive on average in two years out of three until the age of twelve, when fecundity begins to decline. Yeld hinds (those that have failed to rear a calf either because they do not become pregnant in the autumn or because their calf died shortly after birth) are heavier and fatter than milk hinds (those that have reared a calf) and, as a result, are more likely to conceive and do so earlier in the breeding season.

Stags seldom hold harems successfully before they reach adult body weight at between five and six years of age. After the age of ten or eleven, their body condition begins to deteriorate and their fighting ability and reproductive success falls rapidly.

Research on other Scottish deer populations has now shown that the demography of mainland deer populations is similar to that of the Rhum deer (see table 7.1). Perhaps because little of the low ground is fenced for farming, the density of deer on Rhum is higher than in many areas of the West Highlands, but it is not unusual and nor are other measures of growth and reproduction.

Variation in Breeding Success

Though reproductive success changes with age in much the same way in most individuals, some are consistently successful at breeding while others are consistently unsuccessful. Some hinds never rear a calf successfully in their lives while others rear as many as twelve calves in their lifetime. Breeding success varies even more widely among stags, though not as widely as one might suppose given the big differences in harem size: our estimates suggest that the lifetime breeding success of stags ranges from zero to around thirty surviving offspring.

The determinants of breeding success are very different in the two sexes. Among hinds, variation in the number of calves reared over their lifetime is principally caused by differences in calf survival – rather than by differences in the number of calves born. Calf survival is affected by the size of the female's matrilineal group and her dominance rank: animals belonging to large matrilineal groups and subordinate hinds are less likely to rear their calves successfully.

In contrast, the number of calves fathered by stags during their lifetime is determined principally by the size of their harems and the number of hinds they mate with. This depends on their fighting ability and body size – which, in turn, depend on their growth during the first years of their life. Small yearlings are generally small adults and seldom turn into successful stags. A stag's early growth is strongly affected by the mother's milk yield and factors likely to affect this, such as her age, rank and the quality of her home range.

Differences between males and females in the factors affecting breeding success are reflected in the behaviour of mothers to their sons and daughters. Stag calves not only have longer gestation lengths than hind calves but they are born heavier and are allowed to suckle longer and more frequently. This has important costs to the mother: hinds that have reared a stag calf are nearly twice as likely to fail to produce a calf the following season as those that have reared hind calves and if they conceive, do so later in the season. Possibly as a result of these differences in the costs of rearing males and females, the sex ratio at birth varies with the mother's social rank: dominant hinds, who can afford to rear large, heavy calves, produce consistently more sons than subordinates.

Feeding Ecology

Throughout the year, both stags and hinds use the short herb-rich *Agrostis/Festuca* grasslands and the longer species-poor grasslands growing along the streams. The matted *Molinia*-dominated grassland growing on the gentle ridges of the old lazy beds are also used, but to a lesser extent, as is the heather-dominated moorland. The bouncy, white-tussocked *Eriophorum* bog, thick with sphagnum moss and sundew, is little used.

The extent to which the deer use different plant communities changes with the seasonal cycle of plant growth. In spring, when the first growth produces bright green flushes on the short, herb-rich greens behind the beaches, the deer spend much of their time grazing on broad-leaved grasses. On the heather-covered hillsides, they also eat the first green spikes of purple moorgrass (*Molinia*) and deer sedge (*Trichophorum*) but these are quickly abandoned as their leaves lengthen and toughen. Use of the short greens also declines through the summer as the deer gradually strip the available food, nibbling the grass down until it is no more than two or three centimetres high. Instead, they turn to feed on the longer grasslands beside the stream beds and between the dark green clumps of rushes.

As summer progresses, the deer also use heather more and more, searching the upper slopes of the hills for areas where it has been well grazed and is bright green with new growth. In autumn, just before the rut, the herb-rich greens show a second spurt of growth and the deer return to them for a spell. Then, as the days shorten and the hillsides turn brown, they gradually turn again to the longer, coarser swards – *Molinia*-dominated grassland and heather moorland. Together with the longer grasslands and *Juncus* marsh, these provide their staple throughout the winter, while the short greens are pale and bare.

The annual cycle of plant growth is reflected in seasonal changes in body weight and condition. Adults of both sexes decline in condition and weight during the winter months. By March and April, the backbones and pelvises of the poorer animals can be seen through their coarse coats. After the calving season in June, yeld hinds (which are not rearing calves) increase in weight to a peak of around 96 kg (live weight) by the end of the summer while milk hinds (which suffer the energetic expenses of rearing calves) remain approximately constant in weight during the summer months. The seasonal pattern of weight change in stags is similar – with the exception that they lose weight rapidly during the October rut. The average (live) weight of mature stags drops by over 10 kg between mid-September and the end of October from a larder weight of 83 kg to 69 kg though some individuals may lose as much as a fifth of their total body weight during these months.

Population Regulation and Breeding Success

On Rhum, predators are uncommon and deer numbers are limited by food availability. One of the principal aims of current research on Rhum is to understand precisely how food availability limits population density. Do unculled deer populations remain approximately stable or fluctuate violently, as do populations of some rodents? Precisely what aspects of the food supply are limiting? At what time of year? How does food shortage affect fecundity, survival and growth? And which categories of animals are most affected? Answers to such questions are rare because they require detailed data spanning many years. Among British vertebrates, the process of population regulation is reasonably well understood in only one species: the great tit. However, quite different processes may be important in social, polygynous mammals like red deer.

To investigate the natural regulation of population density, all culling in the North Block of the island (the 12 km² of land lying to the north and east of the Kinloch/Kilmory road and including Kilmory Glen) stopped in 1972. The hind population of the Block grew steadily from sixty hinds one year old or over in 1971 to one hundred and seventy nine in 1983. Over the same period, stag numbers remained approximately constant at around one hundred and thirty, but began to decline after 1980 to ninety-seven in 1983.

Hind numbers stabilized because the rate of reproduction declined as population density increased. As density rose above fifteen hinds per km^2, the age at which hinds conceived for the first time changed from two to three years and the proportion of milk hinds calving fell from around eighty per cent to forty per cent. The proportion of calves that died during the summer months did not change, but winter calf mortality increased from around five per cent to around thirty-five per cent of the annual calf crop. As a result, calf/hind ratios in spring fell from fifty calves per hundred hinds to twenty-five calves per hundred hinds.

One variable which was apparently unaffected by the rise in population density was the rate of dispersal. During the whole ten year period, only two hinds dispersed from the study area despite the fact that surrounding areas of the island are maintained at lower population density by culling and food is presumably more abundant there. However, high population density caused hinds occupying ranges at the edge of the study to wander more widely, especially in spring and summer.

Rising density also affected stags. As hind density rose, the average date of conception and the timing of the rut became progressively later. The proportion of male calves and yearlings that died rose more rapidly than that of females, male growth during the first two years of life was depressed and antler size and weight among adult stags fell. High density also led to a reduction in the lifespan of adult stags: the mean age at death of stags reaching three years old fell from fourteen years to ten by 1983 while the mean age at death of hinds remained approximately steady at between twelve and fourteen years. As a result of higher rates of mortality among males, the adult sex ratio gradually changed from favouring stags to favouring hinds.

The tendency for high population density to be associated with a female-biased adult sex ratio is found across mainland red deer populations as well as in other ungulate species, though it was not previously realised that this occurs because stags are more susceptible to food shortage than hinds. This has important implications for deer management since it is often argued that high hind densities are desirable to produce and attract stags. In fact, the reverse appears to be the case – high hind densities are likely to be associated with high stag mortality and reduced antler growth.

The long duration of the Rhum study has also made it possible to investigate the effects of climatic differences between years. The number of animals dying each winter is closely related to how cold are the later months – in years when temperatures are low and rainfall is high between January and March, more animals die than in warmer years. Winter weather also has an important influence on the weights of stags the following autumn – after cold winters, stags shot the following season are lighter than average.

Though research on Rhum has answered many questions about the ecology of red deer there are many that still remain to be answered. Rhum's red deer represent one of the island's most important resources. This is not because they are rare or because the population is unusual but because isolation combined with excellent visibility and public ownership of the land makes Rhum an ideal site for field studies of their ecology and behaviour. Each of the three research projects that have used the Rhum deer population has produced fresh insights into fundamental questions concerning reproduction, population ecology and behaviour which have implications that extend beyond the management of red deer to the management of many other ungulate species. Individual studies inevitably have a finite life as research workers achieve their goals, but future generations of zoologists will have new questions to investigate and new hypotheses to test. To them, as to us, Rhum will offer an open air laboratory with few parallels.

Further Reading

Clutton-Brock, T. H., F. E. Guinness and S. D. Albon (1982) *Red Deer: behavior and ecology of two sexes.* University of Chicago Press and Edinburgh University Press.

Clutton-Brock, T. H., Major, M. & Guinness, F. E. (1985) Population regulation in male and female red deer. *J. Anim. Ecol. 54*, 831-46.

Lincoln, G. A., R. W. Youngson and R. V. Short (1970) The social and sexual behaviour of the red deer stag, *J. Reprod. Fert. Suppt. 11*, 71-103.

Lowe, V. P. W. (1969) Population dynamics of the red deer (*Cervus elaphus* L.) on Rhum, *J. Anim. Ecol. 38*, 425-57.

8

PONIES, CATTLE AND GOATS

I. GORDON (Ponies and Cattle), R. DUNBAR,
D. BUCKLAND AND D. MILLER (Goats)

Introduction

When the Nature Conservancy bought the Isle of Rhum in
1957 they took over the management of a herd of ponies
which had had a long association with the island. Cattle and
sheep had also been present on Rhum in the past, but the
complete stocks of these animals were removed by the grazing tenant when he left in 1957. A herd of cattle were reintroduced to Glen Harris in the early 1970s but there are no
plans, at present, to bring back sheep. A population of feral
goats also inhabited the island.

The Rhum ponies (figure 8.1) are related to the ponies
of the Western Isles of Scotland, which are smaller than the
Highland and Skye ponies, being generally between 13.1
and 13.3 hands in height, with few animals over 14 hands.
They are thickset and powerfully built with long, thick
forelocks reaching from the top of the head down to the
muzzle, long, flowing manes and thick, bushy tails. In the
Rhum stud, there is a wide range of colours including the
traditional mouse-dun, chestnut with silver mane and tail,
bay, black and grey duns and characteristically they have
hazel eyes. Occasionally white ponies are born which are not
favoured by the Nature Conservancy Council and are attri-
buted to poorly chosen stallions used in the 1930s. All the
ponies have a dark eel-stripe down their backs and zebra
stripes on their forelegs which some authors suggest links
them with the primitive breeds of ponies from Northern
Europe.

The cattle reintroduced in 1970 are of the Highland
breed (figure 8.2) which are well adapted to the harsh climate

Figure 8.1. Rhum ponies pastured in Kinloch.

Figure 8.2. Highland cattle in Loch an Dornbac.

and poor grazing characteristics of the island. Highland cattle are characterised by their short, stocky build, long wide-splayed horns, snub muzzles, long, flowing forelocks extending over their eyes and nose, and long, thick, hair coats. The colours of the Rhum herd range from yellow to dark red. Black was the original colour of Highland cattle and it was not until the Victorians were attracted by this quaint breed that selective breeding produced the predominantly red colour found today.

Goats (figure 8.3) are confined almost exclusively to the cliff terraces of the west coast of Rhum, where they are to be found between Dibidil in the south-east and Kilmory in the north. The bulk of the population of two hundred individuals live on the west-facing cliffs between Papadil in the south and Glen Guirdil in the north-west, altogether some 15 km of cliff-line rising steeply 200–400 m above the rocky shoreline (figure 8.4).

History

There is some controversy about the origins of the Rhum ponies. Traditionally, they were believed to be descended from ponies which had been off-loaded by the Spanish Armada in 1588 as the ships tried to escape from the British fleet. However, this is doubtful both because the Admiral of the fleet ordered all livestock to be thrown overboard while the fleet was situated off the east coast of Scotland and because the vast majority of pack animals on board were mules. It is more likely that the ponies originate from Spain or the Western Mediterranean and were transported to the west coast of Scotland along well-established trade routes.

The first record of ponies on Rhum is in 1772 when Thomas Pennant described 'an abundance of mares and a necessary number of stallions. The colts are an article of commerce, but (the inhabitants) never part with the fillies'. In 1775 Dr Johnson mentions that Maclean of Coll, the owner of Rhum, described the ponies as being 'very small, but a breed of eminent beauty'. In the *Old Statistical Account* (1796) it was noted that: 'horses are reared for sale on Rhum only. They are hardy and high mettled though of small size'. Around these times, ponies were traditionally used as saddle and pack animals carrying heavy loads, which makes them suitable for their modern use as stalking ponies.

Figure 8.3. A group of billy goats on boulders along the coast of Rhum. Young billies join these all-male groups at about 6 months old.

Figure 8.4. The precipitous cliffs of Bloodstone, on the north-west of the island, rise to 400 m. Goats feed on the *Agrostis-Festuca* grasslands on the cliff faces and seaweed on the seashore.

When the second Marquis of Salisbury bought Rhum in 1845, the ponies were running wild on the hills as they had been since the Clearances. In 1862, Lord Salisbury removed six or seven ponies, mainly stallions, to his stud at Hatfield before selling Rhum to the Campbells in 1869. The remaining ponies were bought by Lord Salisbury's son, Lord Arthur Cecil, to improve the New Forest breed when the entire stock was sold at auction in Oban by the Campbells just before they sold the island to the Bulloughs in 1883. It was from the New Forest stock that John Bullough was able to acquire Rhum blood (a stallion named Skye and two mares named Fanny and Dolly) to re-establish a stud on the island in 1888 (figure 8.5).

Cattle were most important on Rhum during crofting times when each family had their own black cattle for meat, milk and sale. Black cattle were the progenitors of the contemporary black Highland cattle, with shorter legs and less wide-splayed horns. Thomas Pennant described the system of stock allocation: 'every penny-land is restricted to 28 soums of cattle; one milch cow is reckoned to a soum, or 10 sheep; a horse is reckoned at 2 soums. By this regulation every person is at liberty to make up his soums with whatever species of cattle he pleases. No hay is made on the island nor any provender for winter provisions. The domestic animals support themselves as well as they can on spots of grass reserved for the purpose'. Then, as now, the grazing available to the cattle was poor so that few cattle were kept over winter. The rest were either killed and their meat salted for consumption during the lean months of the year, or were sold to the cattle drovers on the mainland. Cattle numbers were probably at their highest in 1822 when the human population was at its maximum. After this period, the island was given over mainly to the grazing of sheep and the highest number of cattle present on the island this century was in 1946 when the tenant grazed about one hundred animals on the island. The remaining forty cattle and 1,700 sheep were removed from Rhum in 1957.

In 1970 the Nature Conservancy decided to introduce a herd of twenty Highland cows to Glen Harris since research on the vegetation of Rhum showed that, due to reduced grazing pressure, detrimental changes were occurring in the species composition. The new grazing regime with cattle and

Figure 8.5. Fanny with Rhum Bloodstone at foot, 1903.
The ponies are still used to bring deer carcasses off the hill.

red deer had a beneficial effect on the species composition of
the vegetation communities in the Harris area and the original herd has been allowed to increase to around fifty animals.

Pennant records the presence of wild goats on the island
as early as the 1770s, so the origins of the population clearly
predate the departure of the crofters to Canada during the
Clearances. According to the *Old Statistical Account,* Rhum
goat hair was regularly sold in Glasgow for wig-making during the eighteenth century. Later, sporting owners realised
their value for stalking and this led to a number of attempts
to improve the more desirable characters of the stock (notably horn size and shape) by introductions from mainland
herds during the early part of the twentieth century. Their
influence is still evident among the males, where two completely different horn shapes can clearly be seen, the wide-
spreading dorcas-type and the closer-set backward-sweeping
ibex-type. Rhum billies acquired a substantial reputation
among the hunting fraternity in the early years of this century

as a result of the size of their horns and their unusually dense coats (figure 8.3).

After they had purchased the island, the Nature Conservancy Council feared that the goats might be a threat to the island's alpine and coastal flora and the initial management plan recommended a substantial reduction in the goat population. Despite heavy culling over a number of years, the goat population remained approximately constant. Later, it was realised that the coastal vegetation was relatively stable despite goat grazing, so culling was reduced.

Management and Reproduction

Management of the pony herd is mainly directed towards keeping the historical continuity of the Rhum stock intact and maintaining a breed suitable for stalking. The main herd stays out on the hill throughout the year, with only the pregnant mares and foals being pastured in Kinloch during the winter. The stallion is kept at Kinloch away from the rest of the herd except when put to selected mares during the summer. The foals are generally born in Kinloch around the beginning of May and remain with their dams until the age of six months. Colts and any unwanted fillies are shipped off the island in November for sale on the mainland.

The ponies are used as pack animals for carrying deer carcasses off the hill during the stag and hind shooting seasons which extend from the end of July until January. They carry loads of up to 110 kg for long distances on rough and hilly terrain and therefore require to be powerfully built with long necks to give the length of rein required for stalking. Suitable animals are chosen and trained from the age of three when they are introduced to bearing loads by carrying weighted sacks smeared with the blood of culled deer. They are then taken out on to the hills by the ghillies to carry hinds and finally are used to carry the heavier stags.

The herd of breeding cows remain on the hill throughout the year, ranging mainly in the Harris area. During the winter they are regularly fed on a high protein supplement (Rumevite) and in early spring on cattle cobs. From mid-June until early autumn, the herd is moved to fresh grazings in the north-western glens of Guirdil and Shellesder and the bull, normally pastured in Kinloch, is put out to roam with the herd in July. In autumn, the herd is brought back to

Table 8.1. Compositions of the five heft-groups between
the Harris and Giurdil rivers in January 1981.

Heft group	Adult males	Adult females	Kids[a]	Total
Gualainn na Pairce	5	4	2	11
Wreck Bay	16	14	1	31
A' Bhrideanach	4	10	1	15
Camas na h-Atha	6	7	–	13
Bloodstone	8	10	1	19
Total in study area	39	45	5	89

[a] Animals between 0-6 months of age.

Harris glen where they winter. The calves are born over
several weeks from the end of March through until the end
of May, remaining with their mothers until mid-November
when the castrated bullocks and surplus heifers are shipped
off the island.

The goat population is divided into a number of stable
heft-groups, each of which consists of 4–10 females. Since a
daughter tends to adopt the same ranging areas as her
mother, the heft-groups usually consist of closely related
females, together with their dependent young of the year.
Males, once independent of their dam at around six months
of age, are only loosely attached to these heft-groups, and
may spend consecutive winters in the ranges of different
heft-groups. Table 8.1 gives the compositions (as of January
1981) of the five heft-groups whose ranges lie between Harris
and Guirdil, together with the numbers of males then living
in each group's range. Note that the heft-groups vary considerably both in size and sex ratio.

The goat mating system is traditionally considered to be
the epitome of promiscuity. The number of females with
whom a male can mate is limited, in principle, only by the
distances he can cover during the rut. On Rhum, females
come into oestrus over a two to three week period in late
August. Males cluster around a receptive female, vying for
access to mate with her in sporadic fights. Relatively quiet
periods during which the dominant male guards (or 'tends')

the receptive female alternate with periods of frantic activity on a roughly four-hour cycle throughout the day. During the periods of hectic activity, the males chase the female round in circles and the female will often seek safety on the rock faces where the larger and heavier males have difficulty in manoeuvring.

A male goat's competitive ability reaches its peak at around 5–6 years of age when he reaches his maximum weight and physical condition. These males normally have priority of access to females in oestrus, guarding them closely between bouts of mating. Males show a rapid fall-off in competitive ability in old age, and the oldest males in the population (those of 7–8 years of age) usually compete ineffectively. Young males are unable to compete directly for females but can manoeuvre more rapidly on the cliff faces, and are sometimes able to get between the guarding male and the female and to attempt copulations on the run.

Male goats are extraordinarily mobile during the rut: any male may be found anywhere between Kilmory and Dibidil. Not only do the males range widely, but they do so at astonishing speeds. One male took only thirty minutes to cover 5 km on terrain that involved a total climb in altitude of well over 500 m.

Most kids are born during January and February when temperatures are at their lowest, with a second peak in births occurring in August. A double birth peak is unusual in British goat populations, most of which have a single peak during the early spring, and probably reflects the adverse climatic conditions at the latitude of Rhum. This apparently ill-adapted pattern is mainly a consequence of the timing of the rut (in late August) which, with gestation fixed at five months, inevitably results in a peak in births during late January and early February. There is a cline in the timing of the rut among northern latitude populations of feral goats running from July in Norway to October in southern England which correlates well with the latitudinal variation in day-length. Ideally, a spring birth season would provide optimum conditions both for the lactating female and the early growth of the kid.

In the goat population, most male mortality occurs in the autumn months following the August rut. In contrast, female mortality is more evenly spread across the year, with no clear tendency to peak in any one season, but tending to be heavier following the main birth peak in the late winter. Post-morten examination of both male and female carcasses indicates that the animals are generally emaciated and in poor physical condition at the time of death. This is often associated with other contributory conditions including badly worn teeth and 'sand-cracks' in the hoof wall.

Goats are relatively easy to age in the field since distinctive annual rings can be seen on the horns making it possible to determine age-specific birth and death rates. Female fecundity is highest in the middle years and lowest early and late in life. The age at first birth is generally not less than twenty-four months in this population (a little older than is reported for populations in climatically more favourable areas). There are, however, exceptions and a small proportion of females kid at the age of twelve months. The overall rate of reproduction is generally low: only slightly more than fifty per cent of the females give birth during any given year. Twins are rare: only two pairs were observed in forty-nine recorded births.

Age-specific mortality rates are initially high, levelling off before increasing gradually into old age. Slightly more than half of all kids born die before six months of age, and only forty-four per cent survive to their first birthday. Among the older animals, mortality rates are generally higher for males than for females, resulting in a lower life-expectancy for males (roughly 5.0 years versus 7.5 years, respectively, for animals that survive their first six months of life). The oldest animals in the population during our study were eight years old for males and eleven years old for females.

Feeding Behaviour

Cattle, goats and ponies have contrasting methods of digesting their food. Ponies are monogastric (having one stomach) and have a system of food processing whereby food is passed into the single chambered stomach, and then via the small intestine into the enlarged caecum where most of the digestion takes place by microbial fermentation. This form of food digestion is relatively inefficient: only a small percentage of the food taken in is digested and this takes place after the food has passed through the small intestine where most absorption of nutrients occurs. Cattle and goats, on the other hand, have a four-chambered stomach which increases the time during which digestion of food can occur. They also ruminate or chew the cud, breaking down the food into small particles and thereby increasing the surface area on which micro-organisms in the rumen can act. This process increases the amount of digested material available and the digested nutrients can then be absorbed across the wall of the small intestine into the bloodstream. Therefore cattle and goats absorb a relatively larger amount of nutrients from any given amount of food of similar quantity and quality than ponies.

As a result of their need to ruminate, cattle and goats have a different daily pattern of activity from that of ponies. Ponies generally spend fifteen to sixteen hours per day grazing and the rest of their time is spent resting. Grazing time is evenly spaced into six or seven bouts of approximately two hours duration throughout the day and night, with rest periods of one to two hours in between. The ponies' activity pattern varies little throughout the year, the only change being a small increase in the number of grazing bouts per day during the summer. The duration of the ponies' feeding bouts appears to be regulated by the emptying and filling of their stomachs. In winter, they still spend a large proportion of the night feeding, even though the energy expenditure due to heat loss is high.

Cattle graze for approximately ten hours a day in winter. They have one short nocturnal feeding bout, centred around midnight, of one to two hours duration, spending the remainder of the night ruminating the food consumed during the day. Daylight grazing bouts are longer than those of the ponies and last for up to five hours. Grazing time is split up into two distinct periods: a short morning and a longer afternoon grazing bout with a period of rumination in between. In the summer, cattle spend twelve to thirteen hours a day feeding and graze for a greater proportion of the night than in winter. Their diurnal activity pattern is similar to that of the ponies with short periods of feeding between longer periods of ruminating and resting.

The activity patterns of goats are dominated by their need to shelter at night in caves or crannies where air temperature is higher than in the open. Because wind speed increases with altitude, temperatures on the cliff face are lower than on the beach and the goats prefer shelters at beach level to shelters higher up the cliff face. This is reflected in seasonal tendencies to remain on or near the beaches during winter and to range right up on to the mountain tops during the summer. During bad weather, in all seasons, the goats are confined to the lower levels of the cliff-line.

Since sheltering in the sea-caves is incompatible with feeding, the goats do not normally feed during the hours of darkness. During daylight, they spend 60–80 per cent of the available time feeding. In August, the activity budgets are disrupted by the rut: males spend as little as 23 per cent of their time feeding during August, half as much as that spent feeding by females. This results in a substantial drop in energy intake during this period, and males suffer considerable weight-loss as a result. However, they apparently anticipate the August weight-loss by feeding more during the spring and early summer when most of the vegetation growth takes place.

The goats' daily activity pattern follows the same routine throughout the year. They spend the night in caves or other sheltered positions on the beach. At first light, they leave these shelters and begin to graze steadily uphill until mid-afternoon, when they move rapidly back down to feed on the raised beaches until dusk. During the winter, the combination of short day-length and bad weather often keeps the goats on the beaches for most of the day.

Habitat Use

The seasonal pattern of habitat use differs between goats, cattle and ponies. Like other ruminants, cattle lack upper incisors and can only feed on short vegetation by grasping the herbage between the lower incisors and a horny upper pad on their top jaw, tearing off a mouthful at a time. The quantity of grass that they can obtain per bite on these communities is very small when compared with that obtained on the longer grasslands where the tongue is first wrapped around the grass before it is torn off. During the winter, when there is little green matter available, 74 per cent of the cattle's feeding time is spent on the vegetation communities characterised by high biomass but low quality (*Schoenus* fen, *Molinia* grasslands, and herb-rich heath). On these communities they are able to obtain the nutrients necessary to fulfil their energy requirements by consuming a poor quality food in bulk. During the summer, as the standing crop on the *Agrostis-Festuca* swards increase, the cattle gradually spend time feeding on the species-rich grasslands.

In contrast, ponies have both upper and lower sets of incisors and can nip herbage close to the ground. Compared with cattle they are able to obtain a relatively large mouthful of grass even on short vegetation types and are better at selecting plant species and plant parts. They are also able to spend a greater part of the 24 hours feeding, as they do not have the constraints of rumination. As a result, ponies feed at all times of the year on the short-cropped, species-rich and species-poor *Agrostis-Festuca* grasslands even when the biomass is very low. In addition, they spend approximately 25 per cent of their feeding time grazing the *Juncus* marsh areas where the biomass is higher than in the *Agrostis-Festuca* grasslands, but the ratio of green/dead matter is lower. In winter, the ponies rely on their ability to close-crop the short, relatively high quality grasslands and to pass large quantities of food through their less efficient alimentary canal very quickly in order to fulfil their energy requirements. Through the summer months, when there are large quantities of green matter available in all vegetation communities, 19 per cent of the ponies' feeding time is spent grazing on the species-rich *Agrostis-Festuca* grasslands, and 62 per cent of their feeding time is spent grazing the species-poor *Agrostis-Festuca* grassland and *Juncus* marsh, where biomass and quality of the sward is fairly high.

Throughout the year both cattle and ponies feed on seaweed, mainly the stalks of kelp (*Laminaria* sp.). Cattle vary little in the amount of time spent feeding on seaweed (10 per cent in summer, 14 per cent in winter) whereas ponies feed on it almost exclusively in winter when good grazing is sparse (1 per cent in summer, 12 per cent in winter). Kelp is favoured less by ponies because of their inability to efficiently digest the seaweed's structural components, although the nutritive value is quite high.

Goats, being smaller, require absolutely less food than cattle or ponies and can spend more of their feeding time searching the sward for nutritious items. This, combined with their smaller incisor breadth, enables them to select sparsely distributed, good quality food items from poor qual-

Table 8.2. Goat use of the main vegetation communities for feeding in different months of the year. The relative proportion of time spent feeding is indicated by ● (5-9%) and ○ (>9%), based on scan samples of feeding animals.

Community[a]	J	F	M	A	M	J	J	A	S	O	N	D
Beach	○	○	○						○	○	○	○
Dry heaths	○	○	○	○					○	●	●	○
Species-poor greens	○	○	○	○	○	○	○	○	○	○	○	○
Wet heaths				○	●	○	○	○	○			
Species-rich greens					○	○	○	○	○	○	●	
Molinia flush							○					

[a] Information on the species composition of the various communities can be found in Ball (this volume).

ity sward. Though there is a marked seasonal variation in the goats' use of the different vegetation communities (table 8.2), the *Agrostis-Festuca* grasslands (the 'greens') provide 43 per cent of the animals' feeding sites. The rest is made up from the beaches and tidal rocks (14 per cent) and the heath communities (27 per cent). Overall, the goats spend more time feeding on the poorer quality greens which are common on the cliff faces. The richer greens are favoured only during the early summer when productivity is at its highest. During the winter months, the dry heaths (with heathers as the main species eaten) and the beaches (where seaweeds, particularly kelp, are the main foods) are particularly important.

Interspecific Competition

As a result of differences in habitat use, competition for food between goats and red deer is reduced. Though both species appear to feed on similar areas along the cliff-line, deer (particularly hinds) feed predominantly on grasses, and show a marked preference for feeding on the species-rich greens. In addition the goats prefer steeper slopes, make less use of the grass-covered platforms at beach level and spend much more time on the various heath communities on the main cliff face. When conditions deteriorate, the goats, with their more catholic feeding habits, move to habitats and plant communities that the less agile and more selective deer rarely use.

However between cattle, deer and ponies no such mechanisms appear to be operating. At Harris heavy grazing pressure by deer reduces the height of the greens so that cattle cannot consume enough nutrients per bite to make it energetically economical for them to feed on these vegetation communities. Instead, as has been discussed earlier, they fulfil their relatively large dietary requirements by feeding on vegetation communities with a greater standing crop even though they have a reduced nutritive value. Winter grazing by cattle on these *Schoenus*- and *Molinia*-dominated vegetation communities breaks up the deep litter on these swards and increases the quantity of spring growth. Deer exploit this increased production of nutritious green leaves and spend more time grazing on areas grazed by cattle during the previous winter.

Where deer and ponies graze on the same areas as at Kilmory the ponies with their upper and lower incisors are able to nip the grass on the species-rich greens very short. In this way they compete with the deer for the nutritious grass on these greens which results in these areas receiving heavy grazing pressure. This reduces the overall height of the sward and therefore the amount of plant material individuals can consume per bite. In this situation one species interferes with the feeding of another and both suffer as a result.

Because different species utilise different vegetation communities at times when nutrient supplies are at a minimum (i.e. the winter) it seems likely that a higher total biomass of animals can be supported on areas of mixed vegetation where a number of species are grazing together than if the area was occupied by only one species. However, further study of the interactions between species of herbivores is needed to determine the optimum combinations for facilitation and minimal competition.

Further Reading

Alexander, P. (1969) Ponies on Rhum, *South of the Border*, May, 16-19.

Boyd, I. L. (1981) Population changes and the distribution of a herd of feral goats (*Capra* sp.) on Rhum, Inner Hebrides, 1960-1978. *J. Zoology 193*, 287-304.

Clutton-Brock, T. H., Greenwood, P. J. and Powell, P. R. (1976) Ranks and relationships in Highland Ponies and Highland Cows. *Zeitschrift für Tierpsychology 41*, 202-16.

Gordon, I. J. (1986) *Feeding Strategies of Ungulates on a Scottish Moorland*. PhD dissertation, University of Cambridge.

Hardie, P. A. (1969) Rhum ponies and their future, *South of the Border*, October, 21-2.

Mottram, J. E. (1978) The Rhum Fold. *Highland Breeders Journal 21*, 38-9.

Russel, V. (1983) Adaptable Highland Ponies. *The Living Countryside 95*, 1888-91.

CHECK LISTS

Below we list the plants and vertebrates recorded on Rhum.
Check lists of invertebrates are too extensive to include here
but the principal sources are given.

Plants

Eggeling, W. J. (1965) Check list of the plants of Rhum,
Inner Hebrides. Part 1. Stoneworts, ferns and flowering
plants. Part 2. Lichens, liverworts and mosses. *Trans.
Bot. Soc. Edinb.* 40, 20-59 and 60-99.

Gilbert, O-C. (1982) Lichen surveys of Rhum Nature Conservancy Council, Edinburgh.

Watling, R. (1969) Check list of the plants of Rhum, Inner
Hebrides. Part 3. Fungi. *Trans. Bot Soc. Edinb.* 40,
497-535.

Native Higher Plants of Rhum
An updated and shortened version of Eggeling (1965) with
additions by Stirling, McCallum-Webster, Fremlin, Page,
Ettlinger and Nature Conservancy Council staff follows.

Charophyta
CHARACEAE
Chara delicatula Agardh: Stonewort. Rare.
Nitella opaca Nees: Stonewort. Rare.
N. translucens Agardh: Stonewort. Rare.

Pteridophyta
LYCOPODIACEAE
Diphasiastrum alpinum L.: Alpine Clubmoss. Locally com-
mon.
L. clavatum L.: Stagshorn Moss. Rare.
Huperzia selago L.: Fir Clubmoss. Widely distributed.
SELAGINELLACEAE
Selaginella selaginoides (L.) Link: Lesser Clubmoss. Not
uncommon.
ISOETACEAE
[*Isoetes echinospora* Durieu: Quillwort. Requires confirma-
tion.]
I. lacustris L.: Quillwort. Locally plentiful.

EQUISETACEAE

Equisetum arvense L.: Common Horsetail. Common.

E. fluviatile L.: Water Horsetail. Locally plentiful.

E. hyemale L.: Dutch Rush. Local.

E. × *dycei* C. N. Page: Hybrid Water Horsetail. Rare.

E. × *trachyodon* A.Br.: Hybrid Dutch Rush. Not uncommon.

E. × *litorale* Kuhlew. ex Rupr.: Frequent.

E. palustre L.: Marsh Horsetail. Fairly frequent.

E. pratense Ehrh.: Shady Horsetail. Very local.

E. × *rothmaleri* C. N. Page: Rare.

E. sylvaticum L.: Wood Horsetail. Occasional.

E. telmateia Ehrh.: Great Horsetail. Rare.

E. variegatum Schleich. ex Web. & Mohr: Variegated Horsetail. Rare.

OSMUNDACEAE

Osmunda regalis L.: Royal Fern. Widely distributed.

HYMENOPHYLLACEAE

Hymenophyllum tunbrigense (L.) Sm.: Tunbridge Filmy Fern. Very rare.

H. wilsonii Hook.: Wilson's Filmy Fern. Locally plentiful.

POLYPODIACEAE

Asplenium adiantum-nigrum L.: Black Spleenwort. Not uncommon.

A. marinum L.: Sea Spleenwort. Abundant.

A. ruta-muraria L.: Wall Rue. Local, nowhere plentiful.

A. septentrionale (L.) Hoffm.: Forked Spleenwort. Very rare.

A. trichomanes L.: Maidenhair Spleenwort. Common.

A. viride Huds.: Green Spleenwort. Local, nowhere plentiful.

Athyrium filix-femina (L.) Roth: Lady Fern. Widespread.

Blechnum spicant (L.) Roth: Hard Fern. Sparingly distributed.

Cryptogramma crispa (L.) R.Br. ex Hook.: Parsley Fern. Surprisingly rare.

Cystopteris fragilis (L.) Bernh.: Brittle Bladder Fern. Uncommon.

Dryopteris aemula (Ait.) O. Kuntze: Hay-scented Buckler Fern. Locally abundant.

D. affinis (Lowe) Fras.-Jenk (*D. borreri*): Golden Scaled Fern. Local.

D. oreades Fomin (*D. abbreviata*): Dwarf Male Fern. Rare.

D. carthusiana (Villar) H. P. Fuchs: Narrow Buckler Fern. Fairly common.

D. dilatata (Hoffm.) A. Gray: Broad Buckler Fern. Fairly common.

D. expansa (C. Presl.) Fras.-Jenk: Alpine Buckler Fern. Rare.

D. filix-mas (L.) Schott: Male Fern. Not uncommon.

Gymnocarpium dryopteris (L.) Newm.: Oak Fern. Very rare.

Oreopteris limbosperma (All.) Holub.: Mountain Fern. Common.

Phegopteris connectilis (Michx.) Watt: Beech Fern. Rare.

Phyllitis scolopendrium (L.) Newm.: Hart's-tongue Fern. Rare.

Polypodium vulgare L.: Polypody. Fairly common.

P. interjectum Shivas: Uncommon.

Polystichum aculeatum (L.) Roth: Hard Shield Fern. Rare.

P. × mantoniae Rothm.: Rare.

Pteridium aquilinum (L.) Kuhn: Bracken. Locally common.

MARSILEACEAE

Pilularia globulifera L.: Pillwort. Rare.

OPHIOGLOSSACEAE

Botrychium lunaria (L.) Sw.: Moonwort. Locally plentiful.

Ophioglossum vulgatum L. ssp. *vulgatum*: Adder's Tongue. Frequent.

Gymnospermae

PINACEAE

Pinus sylvestris L.: Scots Pine. Reintroduced.

CUPRESSACEAE

Juniperus communis L. ssp. *nana* Syme: Dwarf Juniper. Locally abundant.

Angiospermae

Dicotyledones

RANUNCULACEAE

Anemone nemorosa L.: Wood Anemone. Very uncommon.

Caltha palustris L.: Marsh Marigold. Common.

Ranunculus acris L.: Meadow Buttercup. Common.

R. bulbosus L.: Bulbous Buttercup. Plentiful.

R. ficaria L.: Lesser Celandine. Local.

R. flammula L. ssp. *flammula*: Lesser Spearwort. Common.

R. hederaceus L.: Ivy-leaved Water Crowfoot. Sparse.*R. repens* L.: Creeping Buttercup. Locally common.

[*R. sardous* Crantz: Hairy Buttercup. Reported 1884.]

Thalictrum alpinum L.: Alpine Meadow Rue. Not uncommon.

T. minus L.: Lesser Meadow Rue. Local.

Trollius europaeus L.: Globe Flower. Rare.

NYMPHAEACEAE

Nymphaea alba L.: White Waterlily. Common.

PAPAVERACEAE

Papaver dubium agg.: Long-headed Poppy. Not seen recently.

FUMARIACEAE

Fumaria capreolata L.: White Fumitory. Casual.

CRUCIFERAE

Arabidopsis thaliana (L.) Heynh.: Thale Cress. Rare and local.

Arabis hirsuta (L.) Scop.: Hairy Rock Cress. Local.

Barbarea intermedia Boreau.: Medium Winter Cress. Rare.

B. vulgaris R.Br.: Winter Cress. Rare.

Cakile maritima Scop.: Sea Rocket. One record.

Capsella bursa-pastoris (L.) Medic.: Shepherd's Purse. Common.

Cardamine flexuosa With.: Wood Bitter Cress. Common.

C. hirsuta L.: Bitter Cress. Fairly common.

C. pratensis L.: Lady's Smock. Common.

Cardaminopsis petraea (L.) Hiit.: Northern Rock Cress. Widespread.

Cochlearia danica L.: Danish Scurvygrass. Local.

C. officinalis L.: Scurvygrass. Common.

C. scotica Druce: Scottish Scurvygrass. Well distributed.

Draba incana L.: Hoary Whitlow Grass. Sparingly distributed.

Erophila spathulata Láng: Common Whitlow Grass. Abundant.

Raphanus maritimus L.: Sea Radish. Requires confirmation.

R. raphanistrum L.: Wild Radish. Common.

Sinapis arvensis L.: Charlock. Recorded in 1884 and 1964.

Sisymbrium officinale (L.) Scop.: Hedge Mustard. Not seen recently.

Subularia aquatica L.: Awlwort. Rare.

Thlaspi alpestre L.: Alpine Penny Cress. Rare.

VIOLACEAE

Viola canina L. ssp. *canina*: Heath Violet. Fairly widespread.

V. palustris 1 L.: Marsh Violet. Common.

V. riviniana Rchb.: Common Violet. Common.

POLYGALACEAE

Polygala serpyllifolia Hose: Heath Milkwort. Widely distributed.

P. vulgaris L.: Common Milkwort. Common.

HYPERICACEAE

Hypericum androsaemum L.: Tutsan. Occurs sparingly.

H. humifusum L.: Trailing St John's Wort. Recorded 1884 and 1964.

H. perforatum L.: Common St John's Wort. Rare.

H. pulchrum L.: Slender St John's Wort. Abundant.

CARYOPHYLLACEAE

Arenaria norvegica Gunn. ssp. *norvegica*: Norwegian Sandwort. Rare.

A. serpyllifolia L.: Thyme-leaved Sandwort. Well distributed.

Cerastium atrovirens Bab.: Dark Green Mouse-ear Chickweed. Not uncommon.

C. glomeratum Thuill.: Sticky Mouse-ear Chickweed. Common.

C. holosteoides Fr.: Common Mouse-ear Chickweed. Common.

C. semidecandrum L.: Little Mouse-ear Chickweed. Local.

Cherleria sedoides L.: Mossy Cyphel. Common.

Honkenya peploides (L.) Ehrh.: Sea Sandwort. Recorded 1884 and 1964.

Lychnis flos-cuculi L.: Ragged Robin. Plentiful.

Moehringia trinervia (L.) Clairv.: Three-nerved Sandwort. Very rare.

Sagina apetala Ard.: Common Pearlwort. Occurs sparingly.

S. nodosa (L.) Fenzl: Knotted Pearlwort. Well distributed.

S. procumbens L.: Procumbent Pearlwort. Very common.

S. subulata (Sw.) C. Presl: Awl-leaved Pearlwort. Not uncommon.

Silene acaulis (L.) Jacq.: Moss Campion. Plentiful.

S. alba Miller: White Campion. Rare.

S. dioica (L.) Clairv.: Red Campion. Rare and very local.

S. maritima With.: Sea Campion. Local.

[*S. vulgaris* (Moench) Garcke: Bladder Campion. Reported in 1884.]

Spergula arvensis L.: Corn Spurrey. Common.

Spergularia media (L.) C. Presl: Greater Sea Spurrey. Common.

S. rupicola Lebel ex Le Jolis: Cliff Spurrey. Recorded 1964.

Stellaria alsine Grimm: Bog Stitchwort. Widely distributed.

S. graminea L.: Lesser Stitchwort. Recorded from Kinloch woods.

S. holostea L.: Greater Stichwort. Very local.

S. media (L.) Vill.: Chickweed. Abundant.

PORTULACACEAE

Montia fontana L. ssp. *fontana*: Blinks. Fairly plentiful.

CHENOPODIACEAE

Atriplex glabriuscula Edmonst.: Babington's Orache. Local.

A. hastata L.: Hastate Orache. Local.

A. patula L.: Common Orache. Reported 1884.

Chenopodium album L.: Fat Hen. Uncommon.

Salsola kali L.: Prickly Saltwort. Rare.

LINACEAE

L. catharticum L.: Purging Flax. Common and widespread.

GERANIACEAE

Erodium cicutarium (L.) L'Hérit. ssp. *dunense* Andreas: Common Storksbill. Local.

Geranium dissectum L.: Cut-leaved Cranesbill. Occasional.

G. molle L.: Dovesfoot Cranesbill. Local.

G. robertianum L.: Herb Robert. Nowhere common.

OXALIDACEAE

Oxalis acetosella L.: Wood Sorrel. Widespread.

AQUIFOLIACEAE

Ilex aquifolium L.: Holly. Common.

PAPILIONACEAE

Anthyllis vulneraria L.: Kidney Vetch. Abundant.

Lathyrus montanus Bernh.: Tuberous-rooted Bitter Vetch. Common.

L. pratensis L.: Meadow Vetchling. Common.

Lotus corniculatus L.: Birdsfoot Trefoil. Very common.

L. pedunculatus Cav.: Large Birdsfoot Trefoil. Rather rare.

Medicago lupulina: Black Medick. Local.

[*Trifolium arvense* L.: Haresfoot Clover. Reported 1884.]

T. dubium Sibth.: Lesser Yellow Trefoil. Locally common.

T. medium L.: Zigzag Clover. Locally common.

T. pratense L.: Red Clover. Common.

T. repens L.: White Clover. Widespread.

Vicia cracca L.: Tufted Vetch. Not uncommon.

V. orobus DC.: Bitter Vetch. Local.

V. sativa L.: Common Vetch. Uncommon.

V. sepium L.: Bush Vetch. Uncommon.

V. sylvatica L.: Wood Vetch. Rare.

ROSACEAE

Alchemilla alpina L.: Alpine Lady's Mantle. Fairly common.

A. filicaulis Buser: Local.

A. glabra Neyganfind: Well distributed.

A. xanthochlora Rothm.: Rare.

Aphanes arvensis L.: Parsley Piert. Rare.

Crataegus monogyna Jacq.: Hawthorn. Mostly introduced.

Dryas octopetala L.: Mountain Avens. Frequent.

Filipendula ulmaria (L.) Maxim.: Meadowsweet. Local.

Fragaria vesca L.: Wild Strawberry. Widespread.

Geum rivale L.: Water Avens. Widespread.

Potentilla anserina L.: Silverweed. Common.

[*P. crantzii* (Crantz) G. Beck ex Fritisch: Alpine Cinquefoil. Reported 1884. Requires confirmation.]

P. erecta (L.) Räusch.: Tormentil. Abundant everywhere.

P. palustris (L.) Scop.: Marsh Cinquefoil. Locally plentiful.

P. sterilis (L.) Garcke: Barren Strawberry. Rare.
Prunus spinosa L.: Sloe. Rare.
Rosa canina L.: Dog Rose. Nowhere common.
R. pimpinellifolia L.: Burnet Rose. Widely distributed.
R. pimpinellifolia × *sherardii* = *R.* × *gracilis* Woods. Kinloch.
R. sherardii Davies: Sherard's Downy Rose. Kinloch.
R. villosa L.: Downy Rose. Kilmory. Symington Grieve's
  '*R. tomentosa* Sm.' is probably referable here.
Rubus fruticosus L.: Bramble. Locally common. The follow-
  ing microspecies of this aggregate have been recorded:
  R. dasyphyllus (Rogers) Rogers, *R. griffithianus* Rogers,
  R. incurvatus Bab., *R. mucronulatus* Bab., *R. rotundatus*
  P. J. Muell. ex Gener., *R. rotundifolius* (Bab.) Bloxam,
  R. silurum (A. Ley) W. C. R. Wats., *R. sprengelii* Weihe,
  R. selneri, *R. dumnoniensis*.
R. idaeus L.: Wild Raspberry. Common in the Kinloch
  woods.
R. saxatilis L.: Stone Bramble. Widespread.
Sorbus aucuparia L.: Rowan. Widespread.
  GRASSULACEAE
Sedum acre L.: Wall Pepper. Rare.
S. anglicum Huds.: English Stonecrop. Not uncommon.
S. rosea (L.) Scop.: Roseroot. Common.
  SAXIFRAGACEAE
Chrysosplenium oppositifolium L.: Opposite-leaved Golden
  Saxifrage. Locally abundant.
Saxifraga aizoides L.: Yellow Mountain Saxifrage. Rare.
S. hirsuta L.: Kidney Saxifrage. Rare.
S. hypnoides L.: Mossy Saxifrage. Montane, scattered.
S. nivalis L.: Alpine Saxifrage. Rare.
S. oppositifolia L.: Purple Saxifrage. Common.
S. stellaris L.: Starry Saxifrage. Widespread.
  PARNASSIACEAE
Parnassia palustris L.: Grass of Parnassus. Infrequent.
  DROSERACEAE
Drosera anglica Huds.: Great Sundew. Widespread.
D. anglica × *rotundifolia* (= *obovata* Mert. & Koch): Not un-
  common.
D. intermedia Hayne: Long-leaved Sundew. Plentiful.
D. rotundifolia L.: Round-leaved Sundew. Plentiful.
  LYTHRACEAE
Lythrum salicaria L.: Purple Loosestrife. Kinloch.
  ONAGRACEAE
Chamaenerion angustifolium (L.) Scop.: Rosebay Willow-
  herb. Rare.

Circaea lutetiana L.: Common Enchanter's Nightshade. Rare.

[*Epilobium anagallidifoolium* Lam.: Alpine Willowherb. Symington Grieve's '*E. alpinum* L.' is assumed to have been this plant, but there is no other record of it or of the alternative possibility, *E. alsinifolium* Vill. Confirmation is required.]

E. montanum L.: Broad-leaved Willowherb. Common.

E. montanum × *obscurum* (= *aggregatum* Čelak). Kinloch.

E. obscurum Schreb.: Short-fruited Willowherb. Local.

E. obscurum × *palustre* (= *schmidtianum* Rostk.): Rare.

E. palustre L.: Marsh Willowherb. Locally common.

E. parviflorum Schreb.: Hoary Willowherb. Local.

HALORAGACEAE

Myriophyllum alterniflorum DC.: Alternate-flowered Water Milfoil. Quite common.

CALLITRICHACEAE

Callitriche intermedia Hoffm.: Starwort. Not uncommon.

C. stagnalis Scop.: Starwort. Wet places.

ARALIACEAE

Hedera helix L.: Ivy. Well distributted.

HYDROCOTYLACEAE

Hydrocotyle vulgaris L.: Marsh Pennywort. Common.

UMBELLIFERAE

Angelica sylvestris L.: Wild Angelica. Locally common.

Anthriscus sylvestris (L.) Hoffm.: Cow Parsley. Local.

Conopodium majus (Gouan) Loret.: Pignut. Locally plentiful.

Daucus carota L. ssp. *carota*: Wild Carrot. Fairly common.

Heracleum sphondylium L.: Hogweed. Sparingly distributed.

Ligusticum scoticum L.: Lovage. Abundant.

Oenanthe crocata L.: Hemlock Water Dropwort. Local. *O. lachenalii* C. C. Gmel.: Parsley Water Dropwort. Rare.

Sanicula europaea L.: Sanicle. Rare and very local.

EUPHORBIACEAE

Mercurialis perennis L.: Dog's Mercury. Rare.

Euphorbia helioscopa L.: Sun Spurge. Rare.

POLYGONACEAE

Oxyria digyna (L.) Hill: Mountain Sorrel. Frequent.

Polygonum aviculare L.: Knotgrass. Nowhere very plentiful.

P. convolvulus L.: Black Bindweed. No recent record.

P. hydropiper L.: Water Pepper. Common.

P. persicaria L.: Redleg. Local.

P. raii Bab.: Ray's Knotgrass. Recorded in 1961 and 1964.

P. viviparum L.: Alpine Bistort. Rare.

Rumex acetosa L.: Sorrel. Plentiful.

R. acetosella L.: Sheep's Sorrel. Common.

R. crispus L.: Curled Dock. Common.

R. longifolius DC.: Butter Dock. Local.

[*R. maritimus* L.: Golden Dock. Reported by Symington
Grieve in 1894; rejected on distributional grounds.]

R. obtusifolius L. ssp. *obtusifolius*: Broad-leaved Dock. Common.

URTICACEAE

Urtica dioica L.: Stinging Nettle. Plentiful.

U. urens L.: Small Nettle. Waste ground, Kinloch.

MYRICACEAE

Myrica gale L.: Bog Myrtle. Rare.

BETULACEAE

Alnus glutinosa (L.) Gaertn.: Alder. Extinct but re-intro-
duced.

Betula pubescens Ehrh. ssp. *odorata* (Bechst.) E. F. Warb.:
Downy Birch. Widely but sparsely distributed.

CORYLACEAE

Corylus avellana L.: Hazel. Widespread.

Quercus petraea (Mattuschka) Liebl.: Sessile Oak. Uncommon.

Q. robur L.: Pedunculate Oak. Uncommon.

SALICACEAE

Populus tremula L.: Aspen. Frequent.

Salix aurita L.: Eared Sallow. Common.

S. aurita × *repens* (= *S.* × *ambigua* Ehrh.): Local.

S. caprea L.: Goat Willow. Rare.

S. cinerea L. ssp. *atrocinerea* (Brot.) Silva & Sobr.: Common
Sallow. Sparingly distributed.

S. cinerea × *repens* (= *S.* × *varia* Hesl. Harr.): Rare.

S. herbacea L.: Least Willow. Widespread.

S. herbacea × *repens* (= *S.* × *cernua* E. F. Linton): Rare.

[*S. myrsinites* L.: Whortle-leaved Willow. Reported by
Clark, 1956. Confirmation is desirable.]

S. nigricans Sm.: Dark-leaved Willow. Rare.

S. phylicifolia L.: Tea-leaved Willow. Local.

S. repens L.: Creeping Willow. Common.

ERICACEAE

Arctostaphylos uva-ursi (L.) Spreng.: Bearberry. Local.

Calluna vulgaris (L.) Hull: Heather or Ling. Abundant.

Erica cinerea L.: Bell Heather. Plentiful.

E. tetralix L.: Cross-leaved Heather. Common.

Vaccinium myrtillus L.: Blaeberry. Widespread.

V. vitis-idaea L.: Cowberry. Occasional.

PYROLACEAE

Pyrola minor L.: Common Wintergreen. Very rare.

EMPETRACEAE
Empetrum hermaphroditum Hagerup: Hermaphrodite Crowberry. Locally frequent.
E. nigrum L.: Crowberry. Common.

PLUMBAGINACEAE
Armeria maritima (Mill.) Willd. ssp. *maritima*: Sea Pink. Common.

PRIMULACEAE
Anagallis minima (L.) E. H. L. Krause: Chaffweed. Common.
A. tenella (L.) L.: Bog Pimpernel. Frequent.
Glaux maritima L.: Sea Milkwort. Local on the coast.
Lysimachia nemorum L.: Yellow Pimpernel. Rare and local.
Primula vulgaris Huds.: Primrose. Widespread and abundant.

OLEACEAE
Fraxinus excelsior L.: Ash. Kinloch, possibly native.

GENTIANACEAE
Centaurium erythraea Rafn.: Common Centaury. Local.
Gentianella campestris (L.) Börner: Field Gentian. Frequent.

MENYANTHACEAE
Menyanthes trifoliata L.: Bog Bean. Common.

BORAGINACEAE
Anchusa arvensis (L.) Bieb.: Bugloss. Local.
Myosotis arvensis (L.) Hill: Common Forget-me-not. Local.
M. caespitosa K. F. Schultz: Tufted Forget-me-not. Local.
M. discolor Pers.: Yellow and Blue Forget-me-not. Fairly common.
M. scorpioides L.: Water Forget-me-not. Rare.
M. secunda A. Murr.: Creeping Forget-me-not. Plentiful.
[*M. sylvatica* Hoffm.: Wood Forget-me-not. Reported 1884.]
Mertensia maritima L.: Oyster Plant. Rare.

SCROPHULARIACEAE
Digitalis purpurea L.: Foxglove. Widespread.
Euphrasia officinalis L.: Eyebright. Thirteen micro-species. Common.
Melamphyrum pratense L.: Common Cow-wheat. Rare and local.
Odontites verna (Bell.) Dum.: Red Bartsia. Frequent.
Pedicularis palustris L.: Red Rattle. Common.
P. sylvatica L.: Lousewort. Common everywhere.
Rhinanthus borealis (Sterneck) Druce: Boreal Yellow Rattle. Rare.
R. borealis × *minor* (= *R.* × *gardineri* Druce): Recorded from Fionchra – as *R. drummond-hayi* (F. B. White) Druce.

R. minor L.: Yellow Ratle. Widespread; ssp. *stenophyllus*
(Schur.) O. Schwartz and ssp. *monticola* (Sterneck) O.
Schwartz both occur.

Scrophularia nodosa L.: Figwort. Rare.

Veronica agrestis L.: Field Speedwell. Occasional.

V. arvensis L.: Wall Speedwell. Occasional.

V. chamaedrys L.: Germander Speedwell. Common.

V. officinalis L.: Common Speedwell. Common.

V. scutellata L.: Marsh Speedwell. Recorded 1963.

V. serphyllifolia L.: Thyme-leaved Speedwell. Typical plant
well distributed: ssp. *humifusum (Dickson) Syme. Rare*.

OROBANCHACEAE

Orobanche alba Steph.: Thyme Broomrape. Very local.

LENTIBULARIACEAE

Pinguicula lusitanica L.: Pale Butterwort. Uncommon.

P. vulgaris L.: Common Butterwort. Common.

Utricularia intermedia Hayne: Intermediata Bladderwort.
Local.

U. minor L.: Lesser Bladderwort. Local.

U. neglecta Lehm.: Greater Bladderwort. Rare.

LABIATAE

Ajuga pyramidalis L.: Pyramidal Bugle. Widespread.

A. reptans L.: Bugle. Rare.

Galeopsis speciosa Mill.: Large Hemp-nettle. Frequent.

G. tetrahit L.: Common Hemp-nettle. Common.

Lamium amplexicaule L.: Henbit. Not uncommon.

[*L. moluccellifolium* Fr.: Intermediate Dead-nettle. Reported
to have occurred in the past, but not seen recently.]

L. purpureum L.: Red Dead-nettle. Quite widely distributed.

Mentha aquatica L.: Water Mint. Rare.

M. arvensis L.: Corn Mint. Rare.

Prunella vulgaris L.: Self-heal. Abundant.

Scutellaria galericulata L.: Skullcap. Rare.

S. minor Huds.: Lesser Skullcap. Frequent.

Stachys palustris L.: Marsh Woundwort. Common.

S. palustris × *sylvatica* (= *S.* × *ambigua* Sm.): Local.

S. sylvatica L.: Hedge Woundwort. Rare.

Teucrium scorodonia L.: Wood Sage. Frequent.

Thymus drucei Ronn.: Wild Thyme. Common.

PLANTAGINACEAE

Littorella uniflora (L.) Aschers.: Shoreweed. Abundant.

Plantago coronopus L.: Buckshorn Plantain. Frequent.

P. lanceolata L.: Ribwort. Abundant.

P. major L.: Great Plantain. Common.

P. maritima L.: Sea Plantain. General.

CAMPANULACEAE
Campanula rotundifolia L.: Harebell. Recorded 1884 and 1985.

LOBELIACEAE
Lobelia dortmanna L.: Water Lobelia. Common.

RUBIACEAE
Galium aparine L.: Sticky Willie. Not uncommon.
[*G. boreale* L.: Northern Bedstraw. Reported 1884.]
G. palustre L.: Marsh Bedstraw. Widespread and common.
G. saxatile L.: Heath Bedstraw. Abundant and general.
[*G. uliginosum* L.: Fen Bedstraw. Reported 1884.]
G. verum L.: Lady's Bedstraw. Plentiful.
Sherardia arvensis L.: Field Madder. Rare and local.

CAPRIFOLIACEAE
Lonicera periclymenum L.: Honeysuckle. Widespread.

VALERIANACEAE
Valeriana officinalis L.: Valerian. Very rare.
Valerianella locusta L. Betcke: Cornsalad. Rare.

DIPSACACEAE
Succisa pratensis Moench: Devilsbit Scabious. Abundant.

COMPOSITAE
Achillea millefolium L.: Yarrow. Common.
A. ptarmica L.: Sneezewort. Common.
Antennaria dioica (L.) Gaertn.: Mountain Everlasting. Common.
Arctium minus Bernh.: Lesser Burdock. Recorded 1884 and 1985.
Bellis perennis L.: Daisy. Common.
Carlina vulgaris L.: Carline Thistle. Rather uncommon.
Centaurea nigra L. ssp. *nigra*: Hardheads. Nowhere abundant.
Chrysanthemum leucanthemum L.: Gowans. Rare and local.
C. segetum L.: Corn Marigold. Arable weed.
Cirsium arvense (L.) Scop.: Creeping Thistle. Locally common.
[*C. heterophyllum* (L.) Hill: Melancholy Thistle. Reported 1884.]
C. palustre (L.) Scop.: Marsh Thistle. Common.
C. vulgare (Savi) Ten.: Spear Thistle. Common.
Crepis capillaris (L.) Wallr.: Smooth Hawksbeard. Common.
C. paludosa (L.) Moench: Marsh Hawksbeard. Rare.
[*Gnaphalium supinum* L.: Dwarf Cudweed. Reported 1884.]
G. sylvaticum L.: Wood Cudweed. Rare.
G. uliginosum L.: Wayside Cudweed. Nowhere plentiful.
Hieraceum spp.: Hawkweeds. Not uncommon. Eight microspecies recorded.

H. pilosella L.: Mouse-ear Hawkweed. Frequent.
Hypochoeris radicata L.: Cat's Ear. Widespread and common.
Lapsana communis L.: Nipplewort. Rare.
Leontodon autumnalis L.: Autumnal Hawkbit. Frequent.
L. hispidus L.: Rough Hawkbit. Fairly common.
L. taraxacoides (Vill.) Mérat: Hairy Hawkbit. Recorded 1959.
Saussurea alpina (L.) DC.: Alpine Saw-wort. Often abundant.
Senecio aquaticus Hill: Marsh Ragwort. Common.
S. jacobaea L.: Ragwort. Common.
S. vulgaris L.: Groundsel. Common.
Solidago virgaurea L.: Goldenrod. Common.
Sonchus arvensis L.: Field Sow-thistle. Rather rare.
S. asper (L.) Hill: Prickly Sow-thistle. Occasional.
S. oleraceus L.: Smooth Sow-thistle. Occasional.
Taraxacum laevigatum (Willd.) DC.: Red-fruited Dandelion. Not uncommon.
T. officinale Weber: Common Dandelion. Common.
T. palustre (Lyons) DC.: Narrow-leaved Marsh Dandelion. Local.
T. spectabile Dahlst.: Broad-leaved Marsh Dandelion. Local.
Tripleurospermum maritimum (L.) Koch: Scentless Mayweed. Common.
Tussilago farfara L.: Coltsfoot. Rare.
   Monocotyledones
   JUNCAGINACEAE
Triglochin maritima L.: Sea Arrow-grass. Rather rare.
T. palustris L.: Marsh Arrow-grass. Frequent.
   ZOSTERACEAE
Zostera marina L.: Common Eel-grass. Local.
   POTAMOGETONACEAE
Potamogeton alpinus Balb.: Reddish Pondweed. Rare.
P. berchtoldii Fieb.: Small Pondweed. Rare.
P. coloratus Hornem.: Fen Pondweed. Rare.
P. filiformis Pers.: Slender-leaved Pondweed. Rare.
P. gramineus L.: Various-leaved Pondweed. Rare.
P. gramineus × *perfoliatus* (= *P.* × *nitens* Weber): Rare.
P. natans L.: Broad-leaved Pondweed. Rare.
P. pectinatus L.: Fennel-leaved Pondweed. Occasional.
P. perfoliatus L.: Perfoliate Pondweed. Occasional.
P. polygonifolius Pourr.: Bog Pondweed. Abundant.
P. praelongus Wulf.: Long-stalked Pondweed. Occasional.
P. pusillus L.: Lesser Pondweed. Rare.

RUPPIACEAE
Ruppia maritima L.: Tassel Pondweed. Rare.

LILIACEAE
Hyacinthoides non-scriptus L.: Wild Hyacinth. Locally abundant.
Narthecium ossifragum (L.) Huds.: Bog Asphodel. Abundant.
[*Scilla verna* Huds.: Spring Squill. Reported 1884.]
Tofieldia pusilla (Michx.) Pers.: Scottish Asphodel. Locally common.

JUNCACEAE
Juncus acutiflorus Ehrh. ex Hoffm.: Sharp-flowered Rush. Locally common.
J. articulatus L.: Jointed Rush. Common.
J. biglumis L.: Two-flowered Rush. Very rare.
J. bufonius L.: Toad Rush. Plentiful.
J. bulbosus L.: Bulbous Rush. Common.
J. conglomeratus L.: Compact Rush. Abundant.
J. effusus L.: Soft Rush. Frequent.
J. maritimus Lam.: Sea Rush. Rare.
J. gerardii Lois.: Saltmarsh Rush. Locally common.
J. squarrosus L.: Heath Rush. Common.
J. trifidus L.: Three-leaved Rush. Rare.
J. triglumis L.: Three-flowered Rush. Rare and local.
Luzula campestris (L.) DC.: Field Woodrush. Widespread.
L. multiflora (Retz.) Lej.: Many-headed Woodrush. Common.
L. pilosa (L.) Willd.: Hairy Woodrush. Not uncommon.
[*L. spicata* (L.) DC.: Spiked Woodrush. Reported 1884.]
L. sylvatica (Huds.) Gaud.: Greater Woodrush. Well distributed.

AMARYLLIDACEAE
Allium ursinum L.: Ramsons. Widely but sparingly distributed.

IRIDACEAE
Iris pseudacorus L.: Yellow Flag. Common.

ORCHIDACEAE
Coeloglossum viride (L.) Hartm.: Frog Orchid. Well distributed.
Dactylorhiza incarnata (L.) Soó: Meadow Orchid. Local.
D. incarnata × purpurella (= *D. × latirella* (Hall) Soó): Rare.
D. incarnata × maculata : Rare.
D. kerryensis (Wilmott) P. F. Hunt & Summerh.: Rare.
D. maculata (L.) Soó: Heath Spotted Orchid. Widespread, abundant.
D. maculata × purpurella (= *D. × formosa* (T. & T. A. Stephenson) Soó): Rare.

D. purpurella (T. & T. A. Stephenson) Soó: Northern Fen Orchid. Very common in places.
Gymnadenia conopsea (L.) R.Br.: Fragrant Orchid. Frequent.
Hammarbya paludosa (L.) O. Kuntze: Bog Orchid. Rare.
Leucorchis albida (L.) E. Mey: Small White Orchid. Rare.
Listera cordata (L.) R.Br.: Lesser Twayblade. Rare.
Orchis mascula (L.) L.: Early Purple Orchid. Rare.
Platanthera bifolia (L.) Rich.: Lesser Butterfly Orchid. Widespread.
P. chlorantha (Cust.) Rchb.: Greater Butterfly Orchid. Local.

LEMNACEAE

Lemna minor L.: Common Duckweed. Rare.

SPARGANIACEAE

Sparganium angustifolium Michx.: Floating Bur-reed. Frequent.
S. minimum Wallr.: Small Bur-reed. Reported 1955.

CYPERACEAE

Carex arenaria L.: Sand Sedge. Common; dune areas.
C. bigelowii Torr. ex Schwein.: Stiff Sedge. Common.
C. binervis Sm.: Ribbed Sedge. Well distributed, fairly common.
C. caryophyllea Latour.: Spring Sedge. Not uncommon locally.
C. demissa Hornem.L.: Common Yellow Sedge. Nowhere common.
C. dioica L.: Dioecious Sedge. Not uncommon.
C. distans L.: Distant Sedge. Locally common.
C. echinata Murr.: Star Sedge. Abundant.
C. flacca Schreb.: Glaucous Sedge. Widespread; locally common.
C. hostiana DC.: Tawny Sedge. Not uncommon.
C. lasiocarpa Ehrh.: Slender Sedge. Rare.
C. lepidocarpa Tausch.: Long Stalked Yellow Sedge. Rare.
C. limosa L.: Mud Sedge. Uncommon and local.
C. nigra (L.) Reichard: Common Sedge. Plentiful.
C. otrubae Podp.: False Fox Sedge. Rare and local.
C. ovalis Good.: Oval Sedge. Widespread but not very common.
C. pallescens L.: Pale Sedge. Frequent.
C. panicea L.: Carnation Sedge. Widespread and common.
C. pauciflora Lightf.: Few-flowered Sedge. Rare.
C. pilulifera L.: Pill Sedge. Widespread and common.
C. pulicaris L.: Flea Sedge. Common.
C. rostrata Stokes: Beaked Sedge. Locally common.

C. serotina Mérat: Dwarf Yellow Sedge. Rare.

C. sylvatica Hudson: Wood Sedge. Kinloch.

C. vesicaria L.: Bladder Sedge. Infrequent or rare.

Eleocharis multicaulis (Sm.) Sm.: Many-stemmed Spikerush. Locally common.

E. palustris (L.) Roem. & Schult. ssp. *palustris*: Common Spikerush. Common.

E. quinqueflora (F. X. Hartmann) Schwartz: Few-flowered Spikerush. Common.

E. uniglumis (Link) Schult.: Slender Spikerush. Rare.

Blysmus rufus (Hudson) Link: Saltmarsh Flat Sedge. Kinloch.

Eleogiton fluitans (L.) Link: Floating Spikerush. Common.

Eriophorum angustifolium Honck.: Common Cottongrass. Generally distributed.

E. latifolium Hoppe: Broad-leaved Cottongrass. Well distributed.

E. vaginatum L.: Harestail Cottongrass. Locally abundant.

Isoplepis setacea (L.) R.Br.: Bristle Spikerush. Fairly common.

Rhynchospora alba (L.) Vahl: White Beak-Sedge. Common.

R. fusca (L.) Ait. f.: Brown Beak-Sedge. Very rare.

Schoenus nigricans L.: Bog Rush. Widely distributed and common.

Trichophorum caespitosum (L.) Hartm.: Deer Grass. Very abundant.

GRAMINEAE

Agropyron junceiforme (A. & D. Löve) A. & D. Löve: Sand Couch. Uncommon.

A. repens (L.) Beauv.: Couch. Not uncommon.

Agrostis canina L.: Very common.

A. stolonifera L.: Creeping Bent. Plentiful.

A. tenuis Sibth.: Common Bent. Very common.

Aira caryophyllea L.: Silvery Hair-grass. Fairly common.

A. praecox L.: Early Hair-grass. Common.

Alopecurus geniculatus L.: Marsh Foxtail. Common.

A. pratensis L.: Meadow Foxtail. Very rare.

Ammophila arenaria (L.) Link: Marram Grass. Local.

Anthoxanthum odoratum L.: Sweet Vernal Grass. Common.

Arrhenatherum elatius (L.) Beauv. ex J. & C. Presl: False Oat. Common.

Brachypodium sylvaticum (Huds.) Beauv.: Slender False Brome. Local.

Bromus mollis L.: Soft Brome. Weed of arable ground.

Catabrosa aquatica (L.) Beauv.: Water Whorl-grass. Local.

Catapodium marinum (L.) C. E. Hubbard: Darnel Poa. Rare.

Cynosurus cristatus L.: Crested Dogstail. Abundant.

Dactylis glomerata L.: Cocksfoot. Uncommon.

Deschampsia alpina (L.) Roem. & Schult.: Alpine Hair-grass. Rare.

D. caespitosa (L.) Beauv.: Tufted Hair-grass. Frequent.

D. flexuosa (L.) Trin.: Wavy Hair-grass. Plentiful.

Festuca ovina L.: Sheep's Fescue. Uncommon.

F. pratensis Huds.: Meadow Fescue. Exccedingly rare.

F. rubra L.: Red Fescue. Common everywhere.

F. vivipara (L.) Sm.: Viviparous Sheep's Fescue. Abundant.

Glyceria declinata Bréb.: Small Flote-grass. Uncommon.

G. fluitans (L.) R.Br.: Flote-grass. Local.

Helictrotrichon pratense (L.) Pilg.: Meadow Oat. Recorded 1955.

H. pubescens (Huds.) Pilg.: Hairy Oat. Local.

Holcus lanatus L.: Yorkshire Fog. Common.

H. mollis L.: Creeping Soft-grass. Common.

Koeleria cristata (L.) Pers.: Crested Hair-grass. Abundant.

Lolium perenne L. ssp. *perenne*: Perennial Rye-grass. Locally common.

Molinia caerulea (L.) Moench: Purple Moorgrass. Exceedingly abundant.

Nardus stricta L.: Mat Grass. Locally common.

Phalaris arundinacea L.: Reed Grass. Sparingly distributed.

Phragmites communis Trin.: Reed. Common.

Poa alpina L.: Alpine Meadow-grass. Occurs sparingly.

P. annua L.: Annual Meadow-grass. Plentiful.

P. compressa L.: Flattened Meadow-grass. Local.

P. glauca Vahl: Glaucous Meadow-grass. Uncommon.

P. nemoralis L.: Wood Meadow-grass. Uncommon.

P. pratensis L.: Meadow-grass. Common.

P. trivialis L.: Rough Meadow-grass. Frequent.

Puccinellia maritima (Huds.) Parl.: Sea Poa. Very local.

Sieglingia decumbens (L.) Bernh.: Heath Grass. Common.

Trisetum flavescens (L.) Beauv.: Yellow Oat. Recorded 1955.

Invertebrates

Fryer, G. & Forshaw, O. (1979) The Freshwater Crustacea of the Island of Rhum – a faunistic and ecological survey. *Biological Journal of the Linnean Society* 11, 333-67.

Usher, M. B. (1968) Some spiders and harvestmen from Rhum, Scotland. *Bulletin of the British Spider Study Group* 39, 1-6.

Wormell, P. (1982) The Entomology of the Isle of Rhum National Nature Reserve. *Biological Journal of the Linnean Society* 18, 291-401.

Vertebrates

Mammals of Rhum

Red Deer *Cervus elaphus*: Reintroduced. Widespread, naturalised population of about 1,500.

Otter *Lutra lutra*: Widespread along sheltered shores, seldom penetrating far inland.

Brown Rat *Rattus norvegicus*: Widespread and common especially near sea level.

Field Mouse *Apodemus sylvaticus hebridensis*: Widespread and abundant especially at lower altitudes.

Pigmy Shrew *Sorex minutus*: Widespread and common.

Pipistrelle Bat *Pipistrellus pipistrellus*: Common around Kinloch.

Freshwater Fish of Rhum

Salmon *Salmo salar*: Occasional in the Kinloch River where it runs into the sea.

Brown and Sea Trout *S. trutta*: Abundant.

Three-spined Stickleback *Gasterostous aculeatus*: Common in Papadil Loch.

Common Eel *Anguilla anguilla*: Common in the streams.

Reptiles and Amphibians of Rhum

Several notable absentees: There are no snakes, no slow-worms, and neither frogs nor toads. The following have been recorded:

Common Lizard *Lacerta vivipara*: Fairly common and widespread, particularly numerous at Papadil; has been recorded up to about 1,300 ft on Sgurr nan Gillean.

Palmate Newt *Triturus helveticus*: Abundant, particularly in the northern half of the island. Very common in the breeding season in lochans and pools which do not contain trout. Found up to about 2,200 ft on Hallival. Although the Smooth Newt (*T. vulgaris*) was reported on one occasion there has been no sign of its occurrence since, despite special search.

Birds of Rhum

The island's total of birds recorded now stands at 196 (December 1986) although 3 species are now extinct – the Chough, Partridge and Pheasant (the latter 2 had originally been introduced): another species, the White-tailed Sea Eagle, has been reintroduced, while a fifth – the Leach's Petrel – may never have occurred at all. Fifty-six species may be regarded as 'vagrants', having been recorded on 6 or less occasions: 31 of them have been noted only once. Another 49 species are irregular visitors or else occur regularly on pasage but do not breed. Altogether 86 species have nested on the island (not counting failed attempts by the Rook and House Martin, nor the 4 extinct species, nor others, such as Ptarmigan and Corncrake, which bred at one time). The breeding status of the latter, the Storm Petrel, Tree Pipit and Greenshank is not yet clear. Twenty-nine species nest only irregularly, and in this category are several small passerines whose populations are as yet precariously low. However, as the new plantations around Loch Scresort establish, many woodland species will consolidate their numbers and no doubt new ones will colonise. The following check list has been prepared by J. A. Love.

Red-throated Diver *Gavia stellata*: Summer breeder, arriving late March: 10-14 pairs but success variable.

Black-throated Diver *Gavia arctica*: Vagrant; 2 probables early August 1977; 2, 24.3.78.

Great Northern Diver *Gavia immer*: Regular winter visitor (maximum 5); sometimes remaining until June.

White-billed Diver *Gavia adamsii*: Vagrant; one offshore 28/29.5.80.

Little Grebe *Tachybaptus rufficollis*: Vagrant; singles on Loch Scresort 17.10.60; 28/29.10.61; 21.9.61; 18/19.9.70; 6.2.79; 28.8.79.

Great Crested Grebe *Podiceps cristatus*: Vagrant; one Kinloch 14.10.78.

Black-necked Grebe *Podiceps nigricollis*: Vagrant; one Kilmory 15.3.60.

Fulmar *Fulmar glacialis*: Resident breeder, mainly SE coast, slow increase to about 700 pairs.

Sooty Shearwater *Puffinus griseus*: Summer visitor; up to 16 offshore, mostly August – on 7 occasions.

Storm Petrel *Hydrobates pelagicus*: Summer visitor offshore on at least 7 occasions (June to September). Over 50 mist-netted ashore at 3 localities (but no evidence of breeding since claim by Gray in 1871). One retrap from

Northern Ireland which together with a Rhum-ringed bird was recaptured later on the Summer Isles.

[Leach's Petel *Oceanodroma leucorhoa*: Extinct; dubiously claimed lto have bred in 1871.]

Gannet *Sula bassana*: Summer visitor; regular offshore, especially July and August.

Cormorant *Phalacrocorax carbo*: Summer/winter visitor, rarely more than 4 and usually in summer months.

Shag *Phalacrocorax aristotelis*: Resident breeder; number fluctuate 50-200 pairs.

Heron *Ardea cinerea*: Occasional breeder; 1869 (cliff nests), 2-5 (tree nests) in 1960, 1963/64, 1972, 1977/78; up to a dozen on shore, occasionally inland throughout year; one ringed Rhum killed at wires on Skye, April 1966.

Whooper Swan *Cygnus cygnus*: Passage migrant, March/April and October/November. Rarely groups pause on shore or inland lochs.

Pink-footed Goose *Anser brachyrhynchus*: Passage migrant April and late September/October; two have overwintered at Kinloch, and two parties (72 in April and 11 in September) paused briefly in 1979.

White-fronted Goose *Anser albifrons*: Passage migrant April/May and October/November; one (Greenland race) Kinloch October 1957; 17 Kinloch 29.10.57; 5 Kilmory January 1970.

Greylag Goose *Answer answer*: Summer breeder since 1980 but not successful until 1982 (2 broods) and 1983 (3 broods); regular passage migrant April and October.

Bean Goose *Answer fabalis*: Vagrant 1984.

Canada Goose *Branta canadensis*: Vagrant; 12 Kinloch 9.6.66.

Barnacle Goose *Branta leucopsis*: Passage migrant, mostly October/November; one party on 1.5.77; 2 August 1962; 3 on shore 21.8.79.

Brent Goose *Branta bernicla*: Vagrant; 30 Kinloch 18.4.65; 12 (pale-bellied) Kinloch 14.15.10.80.

Shelduck *Tadorna tadorna*: Summer breeder pre 1910, 1959, 1965, 1966, 1970 and regularly thereafter; 1, sometimes 2 pairs.

Wigeon *Anas penelope*: Winter visitor; 1 to 3 October-January, but 20 Kinloch 9.7.72.

Teal *Anas crecca*: Occasional breeder (up to 2 pairs); a few may winter.

Mallard *Anas platyrhynchos*: Resident breeder, a few pairs at inland lochs, up to 20 may overwinter.

Ferruginous Duck *Aythya nyroca*: Vagrant; 1 Kinloch 16.10.79.

Tufted Duck *Aythya fuligula*: Vagrant; 1 Kinloch 31.1.78; 2 Kinloch 15.4.78; pair Long Loch May 1980; 1 Kinloch 7.10.80.

Scaup *Aythya marila*: Vagrant; 1 died Kinloch 11.10.78.

Eider *Somateria mollissima*: Resident breeder since about 1880; up to 12 broods at Kinloch; flocks of up to 60-80 males April and September.

King Eider *Somateria spectabilis*: Vagrant; 1 Kinloch 24.4.66.

Long-tailed Duck *Clangula hyemalis*: Winter visitor; 7 records (maximum 3 birds) mostly December/February.

Common Scoter *Melanitta nigra*: Summer visitor, usually April/May and August.

Surf Scoter *Melanitta perspicillata*: Vagrant; 2 males Kilmory 23.10.77.

Velvet Scoter *Melanitta fusca*: Vagrant; 1 Kinloch 7.10.57; 2 Kilmory 10.7.77.

Goldeneye *Bucephala lclangula*: Winter visitor; 8 records (maximum 5 birds), mostly December/February.

Red-breasted Merganser *Mergus serrator*: Resident breeder; 2 or 3 broods Kinloch. Winter flocks of 10-20.

Goosander *Mergus merganser*: Vagrant; 2 Fiachanius 1980; 1 Kinloch 5.12.81, 3 inland late November 1982.

White-tailed Sea Eagle *Haliaeetus albicilla*: Extinct; breeder before 1909. Since 1975, 62 have been released on Rhum, mostly dispersed.

Hen Harrier *Circus cyaneus*: Passage migrant, mostly April and August; 2 birds summered May-July 1976.

Goshawk *Accipiter gentilis*: Vagrant; 1 Coire Dubh 7.4.71.

Sparrowhawk *Accipiter nisus*: Regular resident breeder since 1977 (previously proved only in 1959, 1968, 1971), but only ll pair; 2 ringed of 1978 brood recovered following winter Skye and Fort William.

Buzzard *Buteo buteo*: Occasional breeder 1950 (2 pairs), 1959 (1 pair), 1977 and 1983; occasional birds overwinter.

Rough-legged Buzzard *Buteo lagopus*: Vagrant; 1 Kinloch 22.5.79.

Golden Eagle *Aquila chrysaetos*: Resident breeder, 3-5 pairs annually (usually 4 pairs) and 1-3 young reared annually.

Osprey *Pandion haliaetus*: Summer visitor since 1970; 7 records (all single birds) May (2); June (2); July (2); October (1).

Kestrel *Falco tinnunculus*: Resident breeder; up to 4 pairs.

Merlin *Falco columbarius*: Resident breeder; 3-4 pairs but fewer in recent years; 1 young ringed 6.7.66, recaptured Tiree 3 months llater.

Peregrine *Falco peregrinus*: Resident breeder ceased in 1970s but since 1980 1 or 2 pairs have attempted to breed again.

Red Grouse *Lagopus lagopus*: Resident breeder though now less numerous than early this century when large numbers reared for shooting.

Ptarmigan *Lagopus mutus*: Vagrant, although formerly bred, until 1820. A pair released in 1888, 7 seen 2 years later, none since until 1 April 1957; 1, 9.7.59; 3, 15.11.69.

Partridge *Perdix perdix*: Introduced 1850s and 1890s; last seen 1909.

Quail *Coturnix coturnix*: Vagrant; 1 Kilmory June 1959; 1 heard 1.6.66, 3+ Kinloch mid-May to 6.7.77.

Pheasant *Phasianus colchicus*: Introduced 1880-1920s, but none shot after 1906, now extinct.

Water Rail *Rallus aquaticus*: Winter visitor October (1 record); November (2); December (1); January (2); February (1); March (1).

Corncrake *Crex crex*: Summer visitor; formerly several pairs bred until 1975; since 1977 1 or 2 heard for brief period each spring.

Moorhen *Gallinula chloropus*: Irregular visitor; single birds at Kinloch April (3 records); August (2); October (1); November (1); December (2).

Coot *Fulica atra*: Vagrant; singles Kinloch 2-9.10.72 and 8.12.73; 3 on loch 3.9.77; 1 Kinloch 28.8.81.

Oystercatcher *Haematopus ostralgeus*: Resident breeder and flocks of up to 30 may overwinter.

Ringed Plover *Charadrius hiaticula*: Summer breeder; about 6 pairs mainly Kilmory area; passage flocks of 10-20.

Dotterell *Charadrius morinellus*: Vagrant; 1 possible 11.7.80; 1, 15.8.80; 11/12.5.81.

Golden Plover *Pluvialis apicaria*: Resident breeder on the hills and flocks of up to 200 may overwinter; nestling ringed in June 1958 found on Tiree on 9.12.59.

Grey Plover *Pluvialis squatarola*: Vagrant; 1 Kinloch 23.1.63; 3 mid-October 1979 (1 until December); 1 Kinloch 19.3.80; 1 Samhnan Insir 10.8.80; 1 Kinloch 9.9.81.

Lapwing *Vanellus vanellus*: Passage migrant, formerly having bred until 1963. Wintering individuals more common in recent years and up to 30 in occasional migrant flocks.

Knot *Calidris canutus*: Vagrant; 1 Kinloch 12.8.58; 4 Kilmory 19.8.65; 3, 29.8.67; 1, 6.4.80; 1, 2.9.80; 2 Kilmory 14.8.82.

Sanderling *Calidris alba*: Passage migrant (parties up to 15) in May (2 records), July (1), August (11), September (1).

Little Stint *Calidris minuta*: Vagrant; 2 Kilmory 28.8.67; 7 Kilmory 19.8.61.

Pectoral Sandpiper *Calidris melanotos*: Vagrant; 1 Kilmory 21/22.5.78.

Curlew Sandpiper *Calidris ferruginea*: Vagrant; 1 Kilmory 14.8.82.

Purple Sandpiper *Calidris maritima*: Vagrant; 8 Rudha na Moine October 1959; 1 Kilmory 31.1.66; 6 Kilmory 13.1.78.

Dunlin *Calidris alpina*: Passage migrant on the shore in flocks of up to 30. May (2 records), June (1), July (2) otherwise August and September. One summered on the hills May 1979.

Jack Snipe *Lymnocryptes minimus*: Winter visitor September-February. Used to feature in game bags early this century.

Snipe *Gallinago gallinago*: Resident breeder and passage migrant. Large numbers shot September-January early this century.

Woodcock *Scolopax rusticola*: Resident breeder in plantations and passage migrant; many shot early this century November-February.

Black-tailed Godwit *Limosa limosa*: Vagrant; 6 Kilmory 3/4.5.58.

Bar-tailed Godwit *Limosa lapponica*: Vagrant; 1 Kilmory 12.8.58; 1 Kinloch late September 1979.

Whimbrel *Numenius phaeopus*: Passage migrant, up to 8 but usually singles; March-December, especially April/May.

Curlew *Numenius arquata*: Resident breeder, slowly increasing; a pair bred from 1960; 2 pairs, 1975; 3 pairs, 1978; and 4 pairs since 1980; up to 30 may overwinter at Harris.

Redshank *Tringa totanus*: Passage migrant; up to 115 especially in autumn; singles throughout the year.

Greenshank *Tringa nebularia*: Migrant; mostly singles but up to 5 throughout the year, especially on passage April and August; a pair possibly bred 1980, 1981.

Green Sandpiper *Tringa ochropus*: Vagrant 1984.

Wood Sandpiper *Tringa glareola*: Vagrant; 1 Kinloch 25/26.8.76.

Common Sandpiper *Actitus hypoleucos*: Summer breeder; arriving from 14 April; most pairs on the coast; occasional small passage flocks.

Turnstone *Arenaria interpres*: Passage migrant; on the coast in flocks of up to 30; 4 records May/June; most August/September; 1 summered on the hills May 1979.

Red-necked Phalarope *Phalaropus lobatus*: Vagrant; 1 Kinloch 5.9.77.

Pomarine Skua *Stercorarius pomarinus*: Vagrant; a few offshore in 1881; 4 off Mallaig 7.5.79.

Arctic Skua *Stercorarius parasiticus*: Summer visitor offshore; up to 4 seen June-September.

Great Skua *Stercorarius skua*: Summer visitor offshore; usually only 1 or 2 June to September.

Little Gull *Larus minutus*: Vagrant; 1 Kinloch 20.3.58; 1 dead Kinloch 20.3.77.

Black-headed Gull *Larus ridibundus*: Summer/winter visitor; singles seen most months of the year; do not breed.

Common gull *Larus canus*: Resident breeder in small colonies both on the coast and inland lochs; small parties on passage.

Lesser Black-backed Gull *Larus fuscus*: Summer breeder; arriving mid-March; up to 70 pairs in 1 or 2 colonies on the east coast.

Herring Gull *Larus argentatus*: Resident breeder; numbers fluctuate – up to 600 pairs may breed on east coast of island; also round west coast.

Iceland Gull *Larus glaucoides*: Vagrant; 1 Kinloch 7.10.71; 1 Kilmory 27.10.79.

Glaucous Gull *Larus hyperboreus*: Vagrant; 1 Kinloch 9.12.78; 1 Kinloch 26.10.80; 1 at sea 13.11.80; 1 Kinloch 18.10.81.

Greater Black-Backed Gull *Larus marinus*: Resident breeder; up to 30 pairs on east coast; also west coast.

Kittiwake *Rissa tridactyla*: Summer breeder; breeding in 1871 and 1934; increased but now fluctuating; 600 ocupied nests (1969) to 1,700 nests (1982); 1,400 in 1983.

Sandwich Tern *Sterna sandvicensis*: Vagrant; singles offshore 3.8.77; mid-June 1979; pair 20.5.79.

Roseate Tern *Sterna dougallii*: Vagrant; 2 probables Samhnan Insir 21.5.64.

Common Tern *Sterna hirundo*: Summer visitor; occasional breeder; about 10 pairs in 1971; 20 pairs in 1972 but no breeding proved subsequently.

Arctic Tern *Sterna paradisaea*: Summer visitor; occasional breeder, in 1774, 1934, 1955-65; a few pairs irregularly in early 1970s; parties up to 70 may feed offshore.

Guillemot *Uria aalge*: Summer breeder; main colonies on south east cliffs and number around 4,000 birds; counts indicate a low increase; about 11% of birds are bridled.

Razorbill *Alca torda*: Summer breeder; more scattered colonies on the SE cliffs and numbers vary 300-700 pairs.

Black Guillemot *Cepphus grylle*: Resident breeder all round the coast; 30-40 pairs on the east side.

Puffin *Fratercula arctica*: Summer breeder in several small colonies but only 50-80 pairs; slight recovery in recent years.

Rock Dove *Columbia livia*: Resident breeder; recorded in 1796 and 1881; about 10-20 pairs but recent slight decrease.

Woodpigeon *Columba palumbus*: Resident breeder since 1955 with 20-30 pairs now around Kinloch; a few pairs in Kilmory/Harris tree plots; flocks up to 250 at Kinloch in winter.

Collared Dove *Streptopelia decaocto*: Resident breeder since 1960; up to 10 pairs in 1976, but only 1 or 2 pairs after hard winters.

Turtle Dove *Streptopelia turtur*: Summer visitor; recently every year in May or June (usually singles though up to 6 in 1979); 1 possible April 1979; 1, 21/22.9.78.

Cuckoo *Cuculus canorus*: Summer breeder throughout the island; arriving usually last week April; parasite mostly Meadow Pipits; depart July/August.

Barn Owl *Tyto alba*: Vagrant; 1 Kinloch 1955; 1 north side 28.5.76.

Tawny Owl *Strix aluco*: Vagrant; 1 suspected Kinloch 1934; 1 reported 1955; 1 seen June 1963.

Long-eared Owl *Asio otus*: Occasional breeder; 1 pair at Kinloch in about 10 seasons during the last 25 years; occasional passage migrant also.

Short-eared Owl *Asio flammeus*: Ocasional breeder; up to 3 pairs Kilmory, Harris and Kinloch in about 8 of the last 25 years.

Nightjar *Caprimulgus europaeus*: Vagrant; 1 Kinloch 10.6.64.

Swift *Apus apus*: Summer visitor; up to 7 seen May to September, mostly June.

Great Spotted Woodpecker *Dendrocopos major*: Vagrant; 1 suspected Papadil 1934; 1 Kinloch 19-21.8.62; 1 Kinloch late September 1972.

Woodlark *Lullula arborea*: Vagrant; 1 Kinloch 26/27.4.78.

Skylark *Alauda arvensis*: Resident breeder, localised; flocks of up to 50 may overwinter.

Sand Martin *Riparia riparia*: Passage migrant; up to 6 being seen April-June.

Swallow *Hirundo rustica*: Summer migrant; occasional breeder at Kinloch 18 times in the last 45 years (2 pairls in 1968; 3 pairs in 1969); pair occasionally breeds Kilmory and/or Harris; arrivels mid-April; flocks of up to 30 on passage.

House Martin *Delichon urbica*: Passage migrant; pair built nest in 1965 and 1979 but did not breed; small numbers April to November; several hundred mid-September 1966.

Tree Pipit *Anthus trivialis*: Passage migrant April-June; usually up to 6 birds; twice heard singing (1958, 1961) but no record of breeding.

Meadow Pipit *Anthus pratensis*: Resident breeder (abundant) and passage migrant (sometimes several hundred spring and autumn).

Rock Pipit *Anthus spinoletta*: Resident breeder on the coast; some autumn passage.

Yellow Wagtail *Motacilla flava*: Vagrant; 1, 27.6.63; 1, 15.9.76; 1, 25.5.77 – all at Kinloch.

Grey Wagtail *Mortacilla cinerea*: Occasional breeder (9 out of the last 25 years); only 1 pair but 2 in 1971.

Pied/White Wagtail *Motacilla alba*: Resident breeder; up to 4 pairs at Kinloch; (1 pair of 'White' Wagtails bred in 1960); birds become scarce in winter.

Waxwing *Bombycilla garrulus*: Passage migrant; usually single but sometimes up to a dozen November-January; only 1 spring record 1/2.4.75; have occurred in 12 out of last 26 winters.

Dipper *Cinclus cinclus*: Resident breeder; 3-5 pairs inland; become scarce in winter; usually near the coast.

Wren *Troglodytes troglodytes*: Resident breeder and widespread; about 25 pairs at Kinloch.

Dunnock *Prunella modularis*: Resident breeder; about 15 pairs at Kinloch and a few pairs Papadil/Kilmory tree plots.

Robin *Erithacus rubecula*: Resident breeder; mostly Kinloch (about 40 pairs); some spring/autumn passage.

Bluethroat *Luscinia svecica*: Vagrant; 1 Kilmory 1.11.77.

Black Redstart *Phoenicurus ochruros*: Vagrant; 1, 26-31.10.69; 1, 9.11.71 and 1, 31.11.71.

Redstart *Phoenicurus phoenicurus*: Passage migrant; occasional breeder (April to June, October (mostly May)); has bred 1960-62.

Whinchat *Saxicola rubetra*: Occasional summer breeder most years; arrive mid-April and up to 6 pairs Kinloch and Kilmory.

Stonechat *Saxicola torquata*: Resident breeder; up to 16 pairs (1974) Kinloch and Kilmory; numbers reduced by hard winters; some autumn passage.

Wheatear *Oenanthe oenanthe*: Summer breeder; arrive from mid-March; widespread; some autumn passage.

Ring Ouzel *Turdus torquatus*: Summer breeder, though numbers reduced from 6-10 pairs in 1934 to 1-4 pairs in recent years; arrive mid-March.

Blackbird *Turdus merula*: Resident breeder; about 40 pairs at Kinloch; a few elsewhere; several hundred on autumn passage; 3 winter ringed birds from Scandinavia; 1 ringed Rhum 27.11.61, recaught Germany 21.3.64, Rhum 1964/65 winter, dead Denmark 10.4.65.

Fieldfare *Turdus pilaris*: Passage migrant; mostly October/November; a few may overwinter; fewer in spring; 2 heard singing at Harris May 1974.

Song Thrush *Turdus philomelos*: Resident breeder; up to 20 pairs at Kinloch; a few elsewhere; common on passage.

Redwing *Turdus iliacus*: Passage migrant; several hundreds October/November; a few may overwinter; fewer on spring passage.

Mistle Thrush *Turdus viscivorus*: Passage migrant and occasional breeder; 1 or 2 pairs now regular at Kinloch.

Grasshopper Warbler *Locustella naevia*: Summer visitor; possibly breeds; 1 or 2 singing at Kinloch 1966, 1969, 1970, 1971, 1978, and 1981.

Sedge Warbler *Acrocephalus schoenobaenus*: Passage migrant and occasional breeder, especially in recent years.

Whitethroat *Sylvia communis*: Summer visitor and occasional breeder.

Garden Warbler *Sylvia born*: Summer visitor; occasional breeder?; singing males at Kinloch 1969, 1970, 1976, 1979, 1980.

Blackcap *Sylvia atricapilla*: Occasional breeder?; irregular visitor; mostly in winter; up to 7 birds (1981); singing males have been heard in 1958, 1960, 1977, 1979, 1980, 1981.

Wood Warbler *Phylloscopus sibilatrix*: Summer visitor; occasional breeder (on at least 8 occasions).

Chiffchaff *Phylloscopus collybita*: Summer visitor from early April; several now breed most years; 9 records over the winter months.

Willow Warbler *Phylloscopus trochilus*: Summer breeder from mid-April; about 40 pairs Kinloch; becoming commoner in young plantations; some autumn passage.

Goldcrest *Regulus regulus*: Resident breeder; about 20 pairs Kinloch; scarcer after hard winters; sometimes an influx October/November.

Firecrest *Regulus ignicapillus*: Vagrant; 1 Kinloch 6/7.11.80.

Spotted Flycatcher *Muscicapa striata*: Summer visitor; up to 4 pairs breed most years; arrive from mid-May.

Red-breasted Flycatcher *Ficedula parva*: Vagrant; 1 Kinloch 10.11.81.

Pied Flycatcher *Ficedula hypoleuca*: Passage migrant; singles but once 3 birds May (3 records); August (2); September (2); October (2); November (1).

Long-tailed Tit *Aegithalos caudatus*: Occasional breeder; numbers (usually only 1 or 2 family parties) may be wiped out by hard winters; periodic influx from the mainland.

Coal Tit *Parus ater*: Resident breeder at Kinloch; (6-10 pairs).

Blue Tit *Parus caeruleus*: Resident breeder (though none nesting 1961-62, 1968/69); up to 10 pairs at Kinloch, possibly increasing slighty; 1 ringed young recovered on Eigg, and another still breeding at 7 years old.

Great Tit *Parus major*: Ocasional breeder; 1 pair 1955, 1969, 1975/76; sometimes 2 pairs each year subsequently.

Treecreeper *Certhia familiaris*: Resident breeder but rarely more than 4 or 5 pairs at Kinloch.

Lesser Grey Shrike *Lanius minor*: Vagrant; 1 Kinloch 10-15.9.71.

Woodchat Shrike *Lanius senator*: Vagrant; 1 Kinloch 28.5.74.

Chough *Pyrrhocorax pyrrhocorax*: Extinct by 1871.

Jackdaw *Corvus monedula*: Winter visitor; up to 5 birds on 10 occasions (September to March); once in June.

Rook *Corvus frugilegus*: Winter visitor September to March; a group tried to breed in the 1930s but nests blown down; 4 Kinloch 30.5.79.

Hooded Crow *Corvus corone*: Resident breeder in moderate numbers, but up to 250 in winter flocks; Carrion Crows are winter visitors only though a nest was found in 1958; bird at Kinloch April/May 1968; another Kinloch 20.4.80.

Raven *Corvus corax*: Resident breeder; several pairs breed and flocks up to 40 in winter.

Starling *Sturnus vulgaris*: Occasional breeder at Kinloch but remains scarce; flocks up to 100 or more sometimes in winter; 3 winter-ringed birds recovered in Norway and Finland.

Rose-coloured Starling *Sturnu roseus*: Vagrant; 1 Kinloch 4-8.5.71.

House Sparrow *Passer domesticus*: Resident breeder at Kinloch; up to 25 pairs; a flock of 100 Kinloch August-December 1963.

Tree Sparrow *Passer* montanus: Winter visitor; 7 records of up to 6 birds (October-February); 20 Kinloch 9.5.62; 6 Harris 1.6.65.

Chaffinch *Fringilla coelebs*: Resident breeder; nearly all Kinloch (50-60 pairs); a few pairs Papadil; up to several hundred in winter, 2 ringed and later recovered on Eigg and on Soay.

Brambling *Fringilla montifringilla*: Winter visitor, mostly November; singles or flocks of up to 50; 1 singing Kinloch May 1977; Rhum-ringed bird recovered on Eigg 15 days later.

Greenfinch *Carduelis chloris*: Occasional breeder; 1 or 2 pairs at Kinloch in 9 out of last 26 years; a few may over-winter.

Goldfinch *Carduelis carduelis*: Winter visitor; up to 8 birds October-April (11 records in all), July (1) and August (2).

Siskin *Carduelis spinus*: Occasional breeder; up to 6 pairs at Kinloch on at least 10 occasions since 1964; previous to that winter visitor only.

Linnet *Carduelis cannabina*: Summer visitor; 1 or 2 seen April (2 records), May (4), June (1); 2 winter records November and March.

Twite *Carduelis flavirostris*: Resident breeder, but not numerous; up to 50 birds on passage,

Redpoll *Carduelis flammea*: Occasional breeder in 10 nyears since 1957; flocks of up to 30 have twice been seen in autumn.

Crossbill *Loxia curvirostra*: Passage migrant; flocks of up to 200 June/July; occasional birds may overwinter; 2 in April 1981.

Scarlet Rosefinch *Carpodacus erythrinus*: Vagrant; 1 Kinloch 11.7.81.

Bullfinch *Pyrrhula pyrrhula*: Winter visitor (maximum 12) and occasional breeder at Kinloch (1 or 2 pairs) on at least 5 occasions.

Lapland Bunting *Calcarius lapponicus*: Vagrant; 3 October 1967.

Snow Bunting *Plectrophenax nivalis*: Winter visitor on high ground most years (September-April) but rarely as many as 50. Twice recorded in the hills in May (1961, 1979).

Yellowhammer *Emberiza citrinella*: Winter visitor; up to 5 birds recorded November-April (16 occasions, mostly March) and once in August.

Rustic Bunting *Emberiza rustica*: Vagrant; 1 Kinloch 30.6.75.

Little Bunting *Emberiza pusilla*: Vagrant; 1 Harris October 1957; 1 Kilmory October 1957; 1 Salisbury's Dam 28.10.75.

Reed Bunting *Emberiza schoeniclus*: Occasional breeder and winter visitor (mainly January and February); up to 3 pairs have bred at Kinloch on at least 5 occasions since 1957.

Red-headed Bunting *Emberiza bruniceps*: Summer visitor April (1 record), May (1), June (1), July (2), August (1), September (1), October (1). Probably all cage bird escapes.

Black-headed Bunting *Emberiza melanocephala*: Vagrant; 1 Kinloch 20.7.75 (probably an escape).

Corn Bunting *Miliaria calandra*: Summer/winter visitor; up to 5 birds have been recorded March (2 records), April (2), May (2), November (1), December (1).

INDEX

Index

Index